ASIAN WATER DEVELOPMENT OUTLOOK 2020

ADVANCING WATER SECURITY ACROSS ASIA AND THE PACIFIC

DECEMBER 2020

ASIAN DEVELOPMENT BANK

Notes:
In this publication, "$" refers to United States dollars.
ADB recognizes "Korea" as the Republic of Korea.
All photos are by ADB.

Cover design by Kookie Trivino.

On the cover: Children play and fetch water from the community tap at the Behala slum area, Kolkata, India
(photo by Amit Verma).

Contents

Tables, Figures, and Boxes

TABLES

FIGURES

BOXES

Foreword by the Asian Development Bank

I am very pleased to introduce the *Asian Water Development Outlook 2020: Advancing Water Security in Asia and the Pacific* (AWDO 2020), the fourth edition of the flagship AWDO series.

Water security plays a fundamental role in inclusive and sustainable economic growth, social well-being, resilience to water-related disasters, and a healthy environment. The need for water security is even more urgent due to the coronavirus disease (COVID-19) pandemic because access to water, sanitation, and hygiene offers the primary line of defense against the spread of COVID-19 and other diseases. However, far too many people across Asia and the Pacific continue to suffer from limited access to these vital services. At the same time, the region has been inundated by water from natural disasters that has ravaged economies and human welfare.

The new AWDO edition serves as a tangible and reliable knowledge resource for ADB's developing members as they address the multifaceted challenges of water security. For example, using updated methodologies and in-depth analysis of water financing and governance developed in partnership with the Organisation for Economic Co-operation and Development (OECD), this edition sheds light on issues that are critical to ensuring water security and equal access for poor and vulnerable groups.

AWDO 2020 reveals how national water security has been improving across Asia and the Pacific since water security was first quantified in AWDO 2013 along the following five key dimensions: (i) rural household water security (water and sanitation); (ii) economic water security (water to sustainably satisfy economic growth); (iii) urban water security (water and sanitation and flood management); (iv) environmental water security (catchment and aquatic health and environmental governance); and (v) water-related disaster security (resilience against droughts, floods, and storms). While many of ADB's developing members are showing substantial progress, national water security still faces challenges due to uncontrolled urbanization and lagging rural development, vulnerability to weather and climate events, and environmental pressures.

As is evident from the case studies in this edition, AWDO has become a reference for water policy reform and investment planning. ADB's developing members are increasingly elevating water security in their development agendas and applying AWDO principles to formulate effective national and subnational water policies. The practical applications, principles, and recommendations of AWDO also align fully with the Sustainable Development Goals and ADB's Strategy 2030.

To address the persistent challenges to achieving water security in the region, ADB continues to develop hard and soft investments that address climate change, support access to water and sanitation, develop the circular economy and efficient water use, scale up nature-based solutions, and target the underprivileged. We are also supporting governance and financial reforms that manage the trade-offs between development and the environment to strengthen water security responsibly as well as to improve the performance of existing assets and mobilize new sources of financing.

AWDO 2020 was developed with invaluable support from our partners, which include the International WaterCentre, the International Water Management Institute, the Korean Institute of Civil Engineering and Building Technology, the OECD, and the Government of Australia. The findings in this edition offer convincing evidence that partnership from all members of society is crucial to strengthening our region's water security, through their much-needed resources and perspectives on capacity building, governance, finance, and infrastructure development.

We are all stakeholders in water security, and I look forward to working together for a more water-secure and resilient Asia and Pacific region.

Masatsugu Asakawa
President
Asian Development Bank

Foreword by the Asia-Pacific Water Forum

The Asian Water Development Outlook (AWDO) 2020, the fourth in the series, is an intellectual product born from the expanded collaboration between the Asian Development Bank (ADB) and its developing members. Spearheaded by ADB and the Asia-Pacific Water Forum, AWDO 2020 produces updated signposts for decision-makers to understand progress made, identify remaining gaps, and, more importantly, guide future investment and finance decisions to achieve the United Nations Sustainable Development Goals of water security in Asia.

Supported generously by the Government of Australia, AWDO 2020 comes at a critical time when a global pandemic has paused the economy and challenged the resilience of the already vulnerable infrastructure, exacerbating weak governance, constrained financial capital, and inadequate capability of many ADB member countries in securing water sustainability. The intensifying climate change impacts further compromise our capacity to respond effectively to the crises.

AWDO is a data-driven and evidence-based report. The 2020 edition shows an improvement of data sets, the evolution of the indices, and depth of knowledge and insights, as conveyed by the sophistication of the assessments through the interplays of the five key dimensions (KDs). Two important crosscutting features—governance and finance—are enhanced by active engagement with reputable partners from the academic world and international organizations. Needless to say, AWDO 2020 displays rigorous analysis and high-standard collaboration.

In the Great Reset era, AWDO 2020 continues to contribute to thought leadership, policy development, and tracking water security progress in Asia. Sustaining water security for 60% of the world population is an arduous but must-accomplish task. The world shares the vision to embrace the transition toward nature positive and net zero carbon emissions by 2050. This vision renews momentum and creates opportunities for the Asia and Pacific nations to focus on building a better future with a solid water-secure foundation.

The next AWDO will demonstrate how quickly and effectively Asia and the Pacific has progressed in water security, as evidenced by data and increasingly more sophisticated and impactful assessments supported by a stronger and more diverse partnership.

The Asia-Pacific Water Forum is privileged to be part of this landmark project.

Ravi Narayanan
Chair
Governing Council, Asia-Pacific Water Forum

Changhua Wu
Vice Chair
Governing Council, Asia-Pacific Water Forum

Foreword by the Organisation for Economic Co-operation and Development

Water security is a foundation of societal resilience, driving inclusive growth and people's well-being. Three billion people with no access to handwashing facilities have been hit particularly hard by the COVID-19 pandemic. This has reemphasized the importance of sound water policies, which feature prominently in Sustainable Development Goals and targets, notably SDG 6 "clean water and sanitation".

Since 2007, the Asian Water Development Outlook has raised the profile of water security in the region's policy agenda; contributing to a policy-relevant definition of water security, developing a metric to measure progress toward water security, and inspiring local and national reforms. Capitalizing on our expertise, evidence-based analyses, and policy standards—especially the OECD Council Recommendation on Water and the OECD Principles on Water Governance—the OECD is delighted to be a key partner with the Asian Development Bank (ADB) and the Asia-Pacific Water Forum for the 2020 edition of the Asian Water Development Outlook.

The economic and social case for investing in water security in the Asia and Pacific region has been well established. However, water investments come at a distinctly high cost to keep pace with rapid urbanization and economic development, a changing climate, concomitant health, economic, social and environmental crises, and inequality. Such costs can be minimized and covered through mechanisms enabling investment in the future to deliver long-term benefits, combined with effective governance.

Financing water security in Asia and the Pacific is not solely about money. It is also about effective and efficient use of available water resources and assets, as well as ensuring that ongoing expenditures deliver substantial benefits for communities, the economy and the environment. In this regard, promising developments of regional and global relevance were discussed at the regional meeting of the OECD Roundtable on Financing Water, co-organised with ADB in November 2019.

Effective governance is essential to create synergies and trust, as well as to manage complexity. It is key to minimizing fragmentation, catalyzing needed finance, fostering basin management, strengthening capacity, engaging stakeholders, fighting corruption, and assessing whether existing institutions deliver intended outcomes. Insights from the 14th OECD Water Governance Initiative meeting, in November 2020, supported the findings of this report.

I trust that the Asian Water Development Outlook 2020, combined with the excellent and long-lasting collaboration on water policy between ADB and the OECD, will guide future decisions that contribute to better water policies for better lives in Asia and the Pacific.

Angel Gurria
Secretary-General
Organisation for Economic Co-operation and Development

Messages

Because water plays a critical role in all aspects of life, achieving its security is foundational to our communities' safety and prosperity. This is why the Asian Development Bank (ADB) has spent over a decade working to understand and enlighten the world through the Asian Water Development Outlook (AWDO) series.

The AWDO series has been a boon to global development. Over the past decade, Asia and Pacific economies have seen unprecedented growth, accounting for over two-thirds of global economic growth in 2019 alone. However, the coronavirus disease 2019 (COVID-19) pandemic has plunged the region—and the world—into an economic crisis vastly different from any in recent memory. Unlike the 1997 Asian financial crisis and the 2008 global financial crisis, the dual health and economic crisis we are experiencing in 2020 will require more than monetary policy in the long term. According to the United Nations, a global population of 2.2 billion lack access to drinking water and 4.2 billion to basic sanitation. Water security must, therefore, be central to recovery efforts, linking health, safety, and economic growth. Now more than ever, high-quality water and sanitation services are necessary to restart economies.

With global development and rapid urbanization, we see not only great prosperity but also enhanced risks. This report shows that sustainable, circular economies are needed to continue growth and mitigate climate vulnerabilities. By embracing Green Growth, our leaders can harness the post-COVID-19 economic recovery to decarbonize economies and build climate-resilient communities. To grasp the opportunity this growth represents, we must design green, healthy, climate-resilient, and inclusive cities, as pointed out by the Global Green Growth Institute.

In July 2020, the Sustainable Development Goal (SDG) 6 Global Acceleration Framework was launched as a milestone in the United Nations Water Action Decade 2018–2028. ADB's Strategy 2030 reinforces these efforts by prioritizing water security investments. Asia and Pacific nations must work together and follow ADB's admirable lead to achieve our SDGs. This AWDO report focuses on the connection between water security, safety, and economic growth. We must facilitate a global strategy that encourages societies to develop and act on a strategic agenda for water.

Han Seung-soo
Founding Chair, High-Level Experts and Leaders Panel on Water and Disasters
Chair, ADB Water Advisory Group

Messages

Water security is an attainable goal for Asia and the Pacific—but not a simple one. It will require that all of the countries of the region are able to benefit from good quality water, reliably delivered to the right place at the right time, with the risks of water-related extreme events such as floods and drought well managed. The effort and investment demanded to meet this challenge are substantial, but the rewards for the people of Asia and the Pacific will be enjoyed as more sustainable and resilient, and more inclusive, social, and economic development.

The problems holding back water security in Asia and the Pacific are not new, but they are many and dynamic, and they vary tremendously. In common with elsewhere in the world, water use in parts of the region continues to outstrip the renewable water resources available. With rapid urbanization and the evolution of cities into megacities, there are new pressures on water that development of infrastructure struggles to keep pace with. And, as urbanization unfolds, there is a risk that attention turns away from the depopulating rural areas, exacerbating inequalities. There is also little doubt that the climate crisis will continue to aggravate stresses on water resources in Asia and the Pacific, bring an increased likelihood of extreme events, and alter the ecosystem services that societies depend on.

The Asian Water Development Outlook (AWDO) uses a methodology and approach for gauging the status—and progress achieved or lost—of vital components of water security. It provides all countries in the region, as well as businesses, investors, and civil society, with a benchmark set of metrics that will be invaluable in guiding understanding of where, and on what, efforts to improve water security should be focused. The assessments delivered in this report should now be used to help in prioritizing the critical gaps in each country and then moving quickly to planning and delivering sustainable solutions. Then, the value in the AWDO will be shown to be application of evidence to catalyzing action and investment that drives improvements and progress.

In the CGIAR, and at the International Water Management Institute, which leads CGIAR's research-for-development on water, we are excited by the potential for this latest edition of the innovative AWDO series to motivate improvements in the management of water. With AWDO's assessment of current trends now in hand, the next steps on the path to attaining water security in Asia and the Pacific are becoming clearer. Set against this backdrop, the opportunity now to be grasped is use understanding of where and how we are falling short on water security to help us meet both societal and environmental water needs sustainably, now and in the future.

Claudia Sadoff
Executive Management Team Convenor and Managing Director, Research Delivery and Impact
CGIAR System Organization

Messages

Rapid urbanization, climate change, and population growth are placing increasing demands and pressures on the world's clean, freshwater resources. Water insecurity is increasingly having a negative impact on health, prosperity, and stability in many countries around the world.

It is therefore more important than ever that policy makers have resources which support sound, evidence-based policy decisions to combat water insecurity and the many challenges that flow from it.

The Asian Water Development Outlook (AWDO) is such a tool. Now in its fourth edition and 13th year, AWDO has become one of the region's foremost resources in tracking progress on water security and designing solutions to improve it.

AWDO 2020 provides a comprehensive and rigorous analysis of water security across Asia and the Pacific. As a holistic index that considers not just water volumes but other factors like time, place, quality, and use, it determines a whole-of-society value for water that can help policy makers ensure no one is left behind.

Centered on the Sustainable Development Goals (SDGs), it also makes an important contribution to getting us back "on track" to reaching SDG 6, the Sustainable Development Goal for Water—ensuring access to water and sanitation for all, especially for women and girls.

The rich amount of data and information in the 2020 Outlook has already helped the International WaterCentre and the School of Medicine at Australia's Griffith University to develop a COVID-19 Water Security index. This index helps us better understand the region's vulnerabilities to COVID-19 through a water lens.

As the driest inhabited continent on the planet, Australia has long and deep experience in managing water scarcity. I'm pleased that the Government of Australia has been able to again support AWDO in 2020 by tapping into this expertise through our Australian Water Partnership, the International WaterCentre, Griffith University, and the University of Queensland.

I'd like to warmly congratulate the Asian Development Bank and the Asia-Pacific Water Forum for their longstanding and ongoing leadership and commitment to this valuable resource.

Jamie Isbister
Australia's Ambassador for the Environment
First Assistant Secretary, Economic Growth and Sustainability Division
Australian Department of Foreign Affairs and Trade

Messages

Understanding and reducing water risk is at the core of water security. As per the Emergency Events Database (EM-DAT), fatalities in Asia and the Pacific due to water-related disasters were reduced by 67% during 2011–2020, compared with the previous decade. This tremendous jump in water security illustrates the region making strides since the turn of the century. This Asian Water Development Outlook (AWDO) 2020 shows that advances in international policy and infrastructure played a vital role in mitigating water risks.

While most nations are growing more resilient to disasters, tens of millions of people are still displaced every year by sudden-onset disasters. According to United Nations Water, annual damages from water-related disasters cost the global economy hundreds of billions of dollars. To strengthen national commitments to building resilience in 2016, the United Nations High-Level Panel on Water released an action plan highlighting water-related disaster risk reduction and management as one of its key priorities. Enhanced disaster-preventative infrastructure is needed across Asia and the Pacific to accomplish this priority. The governments must invest in such capital-intensive public infrastructure to save lives and safeguard economies.

Significant technical advances have been made over the past decades in lifesaving information and communication technology and smart technologies that can anticipate and manage disasters. However, nations continue to underinvest in disaster risk reduction and management research and infrastructure. To enhance regional disaster resilience, leaders must shift their focus from disaster response to disaster preparedness.

This groundbreaking report seeks to understand how sustainable development, international collaboration, and enhanced technology reduce the impacts of natural disasters like flooding, typhoons, and drought. Moving forward, our global community should build on research and invest in risk reduction systems to see further progress toward water security in Asia and the Pacific.

Han Seung Heon
President
Korea Institute of Civil Engineering and Building Technology

Acknowledgments

The Asian Development Bank (ADB) gratefully acknowledges the authors of the Asian Water Development Outlook (AWDO) 2020: Tom Panella (ADB), Coral Fernandez Illescas (ADB), Silvia Cardascia (ADB), and Eelco van Beek (ADB consultant). ADB recognizes the guidance of Ravi Narayan, Asia-Pacific Water Forum Governing Council, throughout the drafting process.

ADB would also like to recognize the efforts of the writers and contributors to the individual key dimension (KD) reports and the Organisation for Economic Co-operation and Development (OECD) studies on governance and finance, which form the building blocks of AWDO 2020. Their methodologies, detailed analyses, and reviews were fundamental to preparing the overall document.

The KD and OECD reports were written by the following:

- KD1: Rural Household Water Security—Lachlan Guthrie and Declan Hearne (Griffith University and International WaterCentre)
- KD2: Economic Water Security—David Wiberg, Vishnu Prasad Pandey, Chandima Subsasinghe, and Naga Manohar Velpuri (International Water Management Institute)
- KD3: Urban Water Security—Steven Kenway, Julie Allan, William Mogg, Maria Jawad, Beata Sochacka, and Alice Strazzabosco (University of Queensland, Griffith University, and International WaterCentre)
- KD4: Environmental Water Security—Ben Stewart-Koster, Jereme Harte, Christopher Ndehedehe, Kaitlyn O'Mara (Australian Rivers Institute of Griffith University and International WaterCentre); Pennan Chinnasamy (NM Sadguru Foundation and Indian Institute of Technology Bombay); and Pamela Green (City University of New York)
- KD5: Water-Related Disaster Security—Ilpyo Hong, Michael Bak, Jihyeon Park, Hyeonjun Kim, Seongjoon Byeon (Korea Institute of Civil Engineering and Building Technology)
- Governance—Maria Salvetti and Aziza Akhmouch (OECD) and Matthew England (consultant)
- Financing— Xavier Leflaive, Harry Smythe, and Hannah Leckie (OECD)

The following ADB specialists closely guided the KD and OECD reports: Christian Walder (KD1 and KD3), Silvia Cardascia (KD1 and KD3), Sanmugam Prathapar (KD2 and KD4), Coral Fernandez Illescas and Noriyuki Mori (KD2), Stephane Bessadi (KD3), Jelle Beekma (KD4), and Geoffrey Wilson (KD5). ADB also had discussions on AWDO methodology with Avinash Mishra, Advisor to NITI Aayog, India during its development.

ADB is thankful to the following for providing the individual case studies, which enhanced our understanding of water security:

- Thailand: Piyatida Ruangrassamee, Department of Water Resources Engineering, Faculty of Engineering, Chulalongkorn University
- Karnataka, India: Rao Somasekhar Polisetti, Advanced Centre for Integrated Water Resources Management, Water Resources Department, Government of Karnataka

- Timor-Leste: Tiago De Jesus Ribeiro (ADB), and Bronwyn Powell and Lachlan Guthrie (International WaterCentre)
- Yellow River Basin, People's Republic of China: Silvia Cardascia (ADB), Rabindra Osti (ADB), and Sun Yangbo (Yellow River Conservancy Commission)

ADB would also like to thank Woochong Um and Robert Guild (Sustainable Development and Climate Change Department, ADB) for their leadership and guidance, Gino Pascua (ADB consultant) for dedicated effort and patience in preparing the graphics, and Pia Reyes (ADB) and Ellen Pascua (ADB consultant) for keeping us on track with the report.

We greatly appreciate the wisdom and vision of ADB's Water Advisory Group (WAG): Han Seung-Soo (Chair, former Prime Minister of the Republic of Korea); Bambang Susantono (Vice-President for Knowledge Management and Sustainable Development, ADB); Peter Joo Hee Ng (CEO PUB, Singapore); Yumiko Noda (Veolia, Japan); Chairul Tanjung (Indonesia); Kala Vairavamoorthy (Executive Director, International Water Association); Changhua Wu (Vice Chair, Governing Council, Asia-Pacific Water Forum); and Claudia Sadoff (Director General, International Water Management Institute). ADB would like to pay special respects to Isher Judge Ahluwalia, a former WAG member who passed away in September 2020. She was an economist and policy researcher on food and water issues of global importance. She was also a long-time supporter to ADB and a champion of the underprivileged and vulnerable in India and worldwide.

Abbreviations

ADB	Asian Development Bank
AWDO	Asian Water Development Outlook
CASCI	Catchment and Aquatic System Condition Index
COVID-19	coronavirus disease 2019
DALY	disability-adjusted life year
DMC	developing member country
EGI	Environmental Governance Index
FAO	Food and Agriculture Organization of the United Nations
GDP	gross domestic product
GLAAS	Global Analysis and Assessment of Sanitation and Drinking-Water
GNI	gross national income
IWRM	integrated water resources management
JMP	Joint Monitoring Programme
KD	key dimension
NWS	national water security
NWSI	National Water Security Index
OECD	Organisation for Economic Co-operation and Development
SDG	Sustainable Development Goal
WASH	water, sanitation, and hygiene

Executive Summary

Asia has achieved impressive growth in economic and social welfare during the last decades. Good water management and human capital development remain vital to support economic growth and increase overall social well-being in Asia and the Pacific, especially after the coronavirus disease 2019 (COVID-19) pandemic. Despite the achievements in Asia and the Pacific (home to 60% of the world's population), 1.5 billion people living in rural areas and 0.6 billion in urban areas still lack adequate water supply and sanitation. Of the 49 Asian Development Bank (ADB) members from Asia and the Pacific, 27 face serious water constraints on economic development, and 18 are yet to sufficiently protect their inhabitants against water-related disasters.

This Asian Water Development Outlook (AWDO) report describes the water security status in Asia and the Pacific. Water security in AWDO is the availability of adequate water to ensure safe and affordable water supply, inclusive sanitation for all, improved livelihoods, and healthy ecosystems, with reduced water-related risks toward supporting sustainable and resilient rural–urban economies in the Asia and Pacific region. AWDO has been tracking water security in the region since 2013.

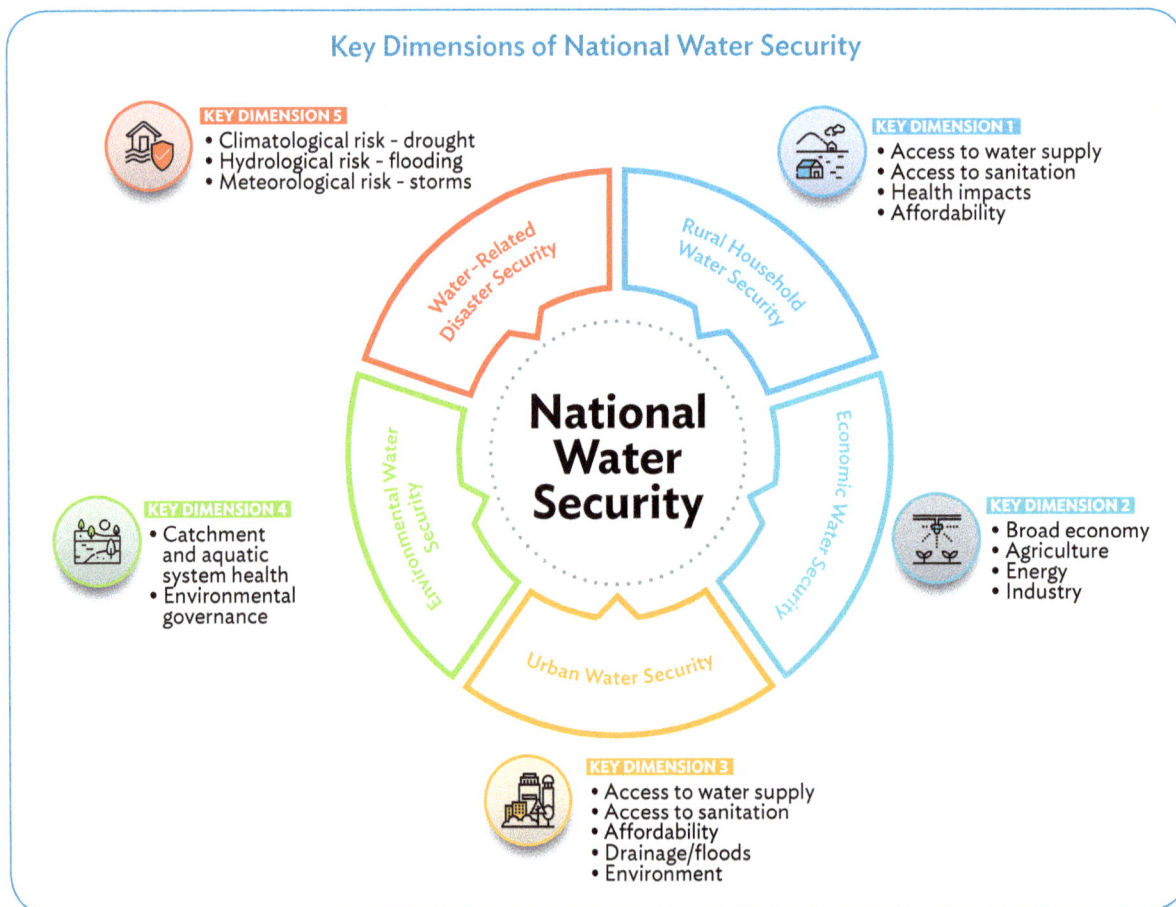

Key Dimensions of National Water Security

KEY DIMENSION 5
- Climatological risk - drought
- Hydrological risk - flooding
- Meteorological risk - storms

Water-Related Disaster Security

KEY DIMENSION 1
- Access to water supply
- Access to sanitation
- Health impacts
- Affordability

Rural Household Water Security

Environmental Water Security

National Water Security

Economic Water Security

KEY DIMENSION 4
- Catchment and aquatic system health
- Environmental governance

KEY DIMENSION 2
- Broad economy
- Agriculture
- Energy
- Industry

Urban Water Security

KEY DIMENSION 3
- Access to water supply
- Access to sanitation
- Affordability
- Drainage/floods
- Environment

In 2007, ADB conceptualized AWDO with the Asia-Pacific Water Forum. The AWDO 2007 report described the need for water security in the region, pointing out that inappropriate management practices, rather than water scarcity, are the main cause of water insecurity. The AWDO 2013 report provided the first quantitative and comprehensive review of water security by using a water security framework with five key dimensions (KDs), as illustrated in the figure—rural household water security, economic water security, urban water security, environmental water security, and water-related disaster security. AWDO 2016 and AWDO 2020 further developed this framework. All KDs are equally important (no weights applied), and the order does not reflect a priority.

Water security is expressed in scores, calculated for each KD based on public data on various indicators describing the KD's performance. The scores of all five KDs are added to form the multidimensional national water security (NWS) score. Banding is applied to these scores to indicate the following NWS and KD development stages: nascent, engaged, capable, effective, and model. None of the 49 ADB members from Asia and the Pacific have achieved the model stage yet, not even the Advanced Economies group. The first two stages (nascent and engaged) place serious constraints on the needed economic and social welfare. Ultimately, ADB members strive to achieve a higher NWS stage. This report presents the results for each ADB member and by the ADB-classified regions: Central

AWDO 2020 Key Messages

Strengthening national water security is a must for improving the quality of life of all people in Asia and the Pacific. Recovering from the setback caused by COVID-19 and adapting to climate change require that all countries put water security at the top of their agendas. Water security enables economic growth and provides the conditions for a healthy and prosperous population. Key recommendations are the following:

- Position water as the centerpiece of sustainable rural development by promoting water-effective irrigation agriculture (KD2), community-based water and sanitation services (KD1), and locally resilient disaster risk reduction (KD5) such as the combination of community protection and farmland flood retention. This will enable a good economic circle of locally affordable investment, income generation, proper management and operation, and an enhanced level of welfare for the people.
- Achieve urban water security (KD3) by investing in water, sanitation, and disaster risk reduction infrastructure (KD5) services not only in cities but also in slums and peripheral areas, while following a gender-based approach.
- Provide a healthy environment (KD4) for the people by drastically reducing pollution, stimulating a circular economy, increasing terrestrial protection, and embracing nature-based solutions for improving water security of other KDs.
- Increase the resilience of the water systems to avoid water-related disasters and to be prepared for climate and other global changes. Turn recent lessons of disasters into better practices of tomorrow by building back better and applying nature-based solutions.

Addressing the above recommendations should be followed with specific attention to women and youth.

It is imperative that countries drastically increase their investment in water, sanitation, and other water-related infrastructure and services by convening all public, private, and innovative financing, which is overwhelmingly lacking, to achieve quality growth and the Sustainable Development Goals in the region. At the same time, financing is needed to enable and sustain a virtuous system of good governance, which requires efficient water-related organizations with sufficient capacity and financial resources to enable them to provide coherent policies, monitor and evaluate progress, and take action when needed, all in interaction with the stakeholders in a transparent way.

and West Asia, East Asia, the Pacific, South Asia, Southeast Asia, and Advanced Economies.

The five KDs of AWDO have strong links with several Sustainable Development Goals (SDGs). Improving water security in each KD has a direct impact on various SDGs. KD1 and KD3 link with target 6.1 (access to safe drinking water) and target 6.2 (access to safe sanitation) of SDG 6 (Clean Water and Sanitation). KD2 contributes to SDG 2 (Zero Hunger) and SDG 7 (Affordable and Clean Energy). KD4 and KD5 link with SDG 3 (Good Health and Well-Being), SDG 14 (Life below Water), and SDG 15 (Life on Land).

Performance and Policy Recommendations

Each KD performance is scored on a scale of 1–20. The scores are presented in the figure below, in which the ADB members are sorted according to their overall score.

National Water Security. Combining the five KDs results in overall water security. No weighting is applied over the five KDs, meaning that a country lagging in one KD might be compensated by its good performance in another KD. Only one economy (Afghanistan) is still in the nascent stage in 2020 (compared with four ADB members in 2013). The number of ADB members in the nascent and engaged stages cumulatively decreased from 30 to 22, while the number of ADB members in the capable and effective stages increased from 19 to 27. There are no ADB members yet in the model stage for the overall NWS. By improving the performance of the five KDs through adequate policies, ADB members can move up from nascent to engaged to capable to effective and ultimately to model. Political choices will determine the priorities in the KDs. During 2013–2020, good progress has been made in KD1 (Rural Household), KD2 (Economic), and KD5 (Disaster), while improvement in KD3 (Urban) and KD4 (Environment) has been slower.

KD1: Rural Household Water Security. All regions show steady progress in improving rural household water security. East Asia and Southeast Asia perform well, while the Pacific and South Asia lag behind. Twenty-three ADB members are still in the nascent and engaged stages (compared with 28 in 2013). The number of ADB members in the effective and model stages has increased from 7 to 13. Increasing rural household water security is paramount in the Asia and Pacific region, given that nearly half of households in the region are still living in rural areas despite urbanization trends. Rural households are more vulnerable, and investing in water and sanitation for rural households is generally less attractive than economic uses like agriculture. AWDO 2020 provides the following policy recommendations related to KD1:

- **Engage vulnerable people in decision-making.** Despite specific policies in some ADB members, the needs of vulnerable people are not being addressed.
- **Invest in human resources capacity.** Human resources appear to be a major constraint on implementing water and sanitation policies. Governments must invest in the human resources required to deliver water services. Special attention should be given to youth and women.
- **Deliver locally appropriate solutions for ADB members in the Pacific.** The Pacific region is lagging, and relatively little progress had been made since 2013. Their specific geographic and financial situation requires tailored approaches.

KD2: Economic Water Security. East Asia has experienced a significant increase in economic water security and has now achieved the same level as Advanced Economies. All other regions also show good progress except the Pacific, where eight ADB members are still in the nascent stage. With 32 ADB members still in the nascent and engaged stages, 2.1 billion people face serious limitations in their economic activities due to insufficient water management. The number of ADB members in the effective

National Water Security Index

1 Nascent   **2** Engaged   **3** Capable   **4** Effective   **5** Model

New Zealand
Japan
Australia
Korea, Republic of
Hong Kong, China
Taipei,China
Singapore
Brunei Darussalam
Malaysia
Kazakhstan
Palau
China, People's Republic of
Kyrgyz Republic
Cook Islands
Armenia
Philippines
Turkmenistan
Azerbaijan
Georgia
Maldives
Samoa
Bhutan
Uzbekistan
Tonga
Mongolia
Indonesia
Niue
Sri Lanka
Viet Nam
Fiji
Thailand
Nauru
Tajikistan
Cambodia
Lao People's Democratic Republic
Tuvalu
Bangladesh
Nepal
Timor-Leste
Vanuatu
Solomon Islands
Marshall Islands
Myanmar
Kiribati
India
Papua New Guinea
Pakistan
Micronesia, Federated States of
Afghanistan

0   10   20   30   40   50   60   70   80   90   100

| Key Dimension 1 | Rural Household | Key Dimension 2 | Economic | Key Dimension 3 | Urban |

| Key Dimension 4 | Environment | Key Dimension 5 | Water-Related Disaster |

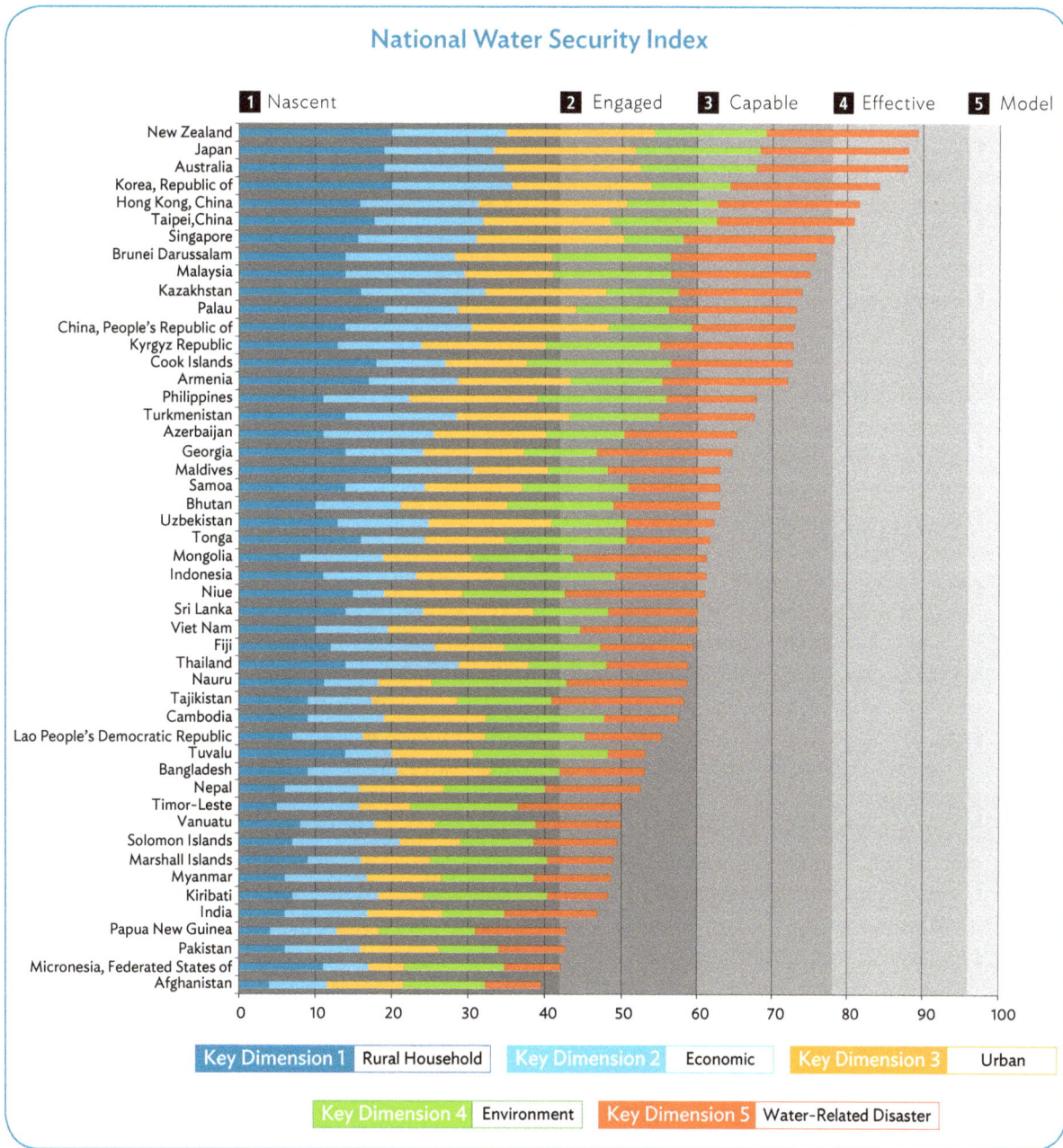

Water Security in Central and West Asia

Water security in Central and West Asia shows reasonable scores for countries of the former Soviet Union but rather weak scores for Afghanistan and Pakistan. The region shows progress in water security during 2013–2020. Growth in rural household water security (KD1) is strong. Economic water security (KD2) has improved during 2013–2016 but decreased during 2016–2020. Urban water security (KD3) remains at the same level and needs attention. Environmental water security (KD4) is low compared with that of the other regions, as is the case for water-related disaster security (KD5). KD5 has shown steady progress during 2013–2020. Priority actions for governance improvement in Central and West Asia are managing trade-offs, integrity, and stakeholder engagement.

stage has increased from three to four. Despite this progress, ADB members need to continue investing in water resource systems that provide the services required to cope with socioeconomic and demographic growth and climate change pressures. The following policy recommendations are relevant for KD2:

- **Enhance water resources monitoring, measurement, and data availability.** Optimizing water resources management requires good data and data availability on supply, demand, quality, benefits, and costs, among other things. Water should be promoted as an economic input to production, and its value accounted for. Data are instrumental in making decisions in investments and developing innovative ways to manage the system.
- **Improve water productivity.** As water becomes increasingly scarce while demand is rising, water should be used as efficiently as possible by promoting robust water allocation regimes, water reuse, conservation practices, and information technology.
- **Ensure adequate storage and distribution mechanism.** This should be done while promoting integrated water resources management (IWRM) and investing in climate change adaptation and resilience. As water availability is often not at the right location and time, water must be stored and transported. Storage, including through wetlands and nature-based solutions, is also necessary to mitigate flood and drought risks and adapt to climate change. An integrated approach is needed to develop and manage the system. Institutional arrangements must account for the integrated nature of water management across sectors and disciplines.

KD3: Urban Water Security. The level of urban water security has remained about the same during 2013–2020. Urban water security in East Asia has nearly reached the same level as the Advanced

Water Security in East Asia

East Asia has shown strong growth in water security during 2013–2020. Taipei,China has the highest water security, but the growth is strongest in the People's Republic of China. Mongolia has also made significant progress. There is room for improvement in rural household water security (KD1) and environmental water security (KD4). Economic water security (KD2) has reached the level of Advanced Economies. Urban water security (KD3) is well taken care of. Water-related disaster security (KD5) has significantly improved since 2013 but needs further growth to reach the level of Advanced Economies. Priority actions for governance improvement in East Asia are policy coherence, managing trade-offs, and stakeholder engagement.

Economies. The Pacific region is lagging. Seven ADB members are still in the nascent stage, while 18 are in the engaged stage, which means that 800 million urban people do not have adequate water supply and sanitation services. Despite the major investments ADB members made, KD3 has remained constant mainly due to rapid urbanization. The rapid growth of urban centers and peri-urban areas, along with climate change impacts, creates significant challenges for the provision of water, wastewater, and stormwater infrastructure. The following policy recommendations are relevant for KD3 to meet these challenges:

- **Invest in sanitation, wastewater treatment, and circular economy.** Combining the challenges in water supply and sanitation might offer possibilities to solve both problems conjunctively, e.g., by direct and indirect potable wastewater recycling and reuse.
- **Improve water cost-effectiveness and affordability.** Affordability has improved over time (e.g., in the Pacific, East Asia, South Asia, and Southeast Asia) or remained stable. Innovation is needed particularly in the Pacific, where affordability is significantly worse than in other regions.

- **Improve drainage security in the Pacific, Central and West Asia, and Southeast Asia.** This improvement requires both an enhanced understanding of the flooding risks in urban areas and larger investments in catchment management, often combining green and gray solutions.

Water Security in the Pacific

Water security in the Pacific region differs by country. Some countries perform well (e.g., Palau and the Cook Islands), but other countries score rather low. The Pacific lags behind other regions in Asia, mainly due to the geographic location and limited financial means. Of particular concern is the low water security in Papua New Guinea and the Federated States of Micronesia. Much effort is needed to improve rural household water security (KD1) and urban water security (KD3). Most Pacific countries score well on environmental water security (KD4) and show resilience in water-related disaster security (KD5). Progress during 2013–2020 is somewhat disappointing. Priority actions for governance improvement in the Pacific are financing, managing trade-offs, and monitoring and evaluation.

KD4: Environmental Water Security. KD4 shows a mixed picture. On the one hand, the number of ADB members in the capable and effective stages has increased from 20 in 2013 to 31 in 2020. On the other hand, 18 ADB members are still in the nascent and engaged stages. Southeast Asia scores well and is nearly at the same level as the Advanced Economies, with the Pacific region scoring above average. South Asia is lagging behind other regions. Restoring and maintaining the health of rivers, wetlands, and groundwater systems is vital for improving KD4 performance and also contributes to other KDs (particularly KD2 and KD3). The following policy recommendations are made to ensure sustainable environmental water security:

- **Improve pollution load management and stimulate circular economic activities.** This includes reducing inefficient agricultural practices, e.g., overuse of fertilizers, leading to the deterioration of water quality in surface and groundwater systems. Treatment and reuse of wastewater from household and industrial systems will reduce the amount of untreated wastewater released into the natural systems while addressing water scarcity.
- **Increase the protection of natural terrestrial and aquatic systems.** This includes the conservation of riparian vegetation, which can filter fertilizers and excess sediment runoff from nearby agricultural operations.
- **Implement measures that make hydrological alteration more sustainable and promote riverine connectivity.** Flow alteration of river and wetland systems is a primary cause of reduced aquatic ecosystem health. ADB members with comparatively poor outcomes for hydrological alteration could be supported to develop locally specific environmental flow programs.
- **Reduce groundwater depletion.** Twenty ADB members received the worst possible rating for groundwater resource sustainability. Solving this issue requires proper aquifer mapping, licensing and compliance enforcement, and regulatory and economic incentive measures for water conservation.

Water Security in South Asia

Water security in South Asia is strongly influenced by the relatively low performance on rural household water security (KD1) and urban water security (KD3). Progress in these dimensions was made during 2013–2020 but not enough to bring this region at the same level as East Asia and Southeast Asia. The region also performs weakly on environmental water security (KD4) but comparatively well on water-related disaster security (KD5). Progress of overall water security was made during 2013–2016 but did not continue during 2016–2020. Priority actions for governance improvement in South Asia are policy coherence, managing trade-offs, and monitoring and evaluation.

KD5: Water-Related Disaster Security. All regions, except South Asia, show good progress in water-related disaster security. South Asia experienced some major disasters in the last 10 years, which influenced their score negatively. In 2020, 18 ADB members are still in the nascent and engaged stages, slightly down from 20 in 2013. Compared with only 11 ADB members in 2013, 19 ADB members are already in the effective and model stages in 2020. Climate change is likely to increase climate variability and the probability of floods and droughts. The following policy recommendations are made to mitigate these risks:

- **Invest in green and gray disaster infrastructure.** Green infrastructure refers to nature-based solutions, while gray infrastructure consists of traditional constructions. Capital-intensive investments are needed to address present problems and mitigate increased risk due to climate change. Public finance for such investment should be paired with alternative financing sources by crowding in commercial finance or climate adaptation financing. Blended financing can help mobilize private sector financial resources.
- **Address gender gaps.** Water-related disasters disproportionately affect marginalized groups, particularly women. More action is needed to analyze and identify gender-specific interventions. Governments should ensure that the most disadvantaged women have access to resilience adaptation programs and funds.
- **Promote integrated flood risk management, including the piloting of nature-based solutions.** Flood risk mapping, integrated land use planning, and early warning systems are cost-effective investments to limit the exposure of people and assets to water-related risk. Investments in gray and green infrastructure and residual risk instruments are complementary measures.
- **Improve data collection, modeling, and associated system for preventive disaster risk management.** As indicated

in KD2 (Economic Water Security), good data are essential for decision-making on investments in disaster risk management. Agencies need to be equipped with modern facilities and techniques for collecting and assessing data.

Water Security in Southeast Asia

Southeast Asia shows promising steady progress in water security during 2013–2020, with Malaysia as the top performer. The region also shows a balanced performance of all five key dimensions (KDs). Strong progress is seen in rural household water security (KD1), and other KDs show steady progress. The environmental water security (KD4) is about at the same level as the Advanced Economies. Priority actions for governance improvement in Southeast Asia are policy coherence, integrity, and stakeholder engagement.

Governance and Financing Needs

Achieving water security for all ADB members in Asia and the Pacific requires robust public policies across all levels of government, a clear allocation of duties among responsible authorities, as well as regular monitoring and evaluation mechanisms. At the same time, adequate financial means should be made available to implement the needed institutional and technical interventions, which require appropriate enabling environments to make the best use of available water resources, water-related assets, and financial sources.

Governance. In AWDO 2020, the Organisation for Economic Co-operation and Development (OECD) has surveyed 48 ADB Asia and Pacific members using the 12 OECD Principles on Water Governance to shed some light on the governance gaps in the region.* Based on the survey's outcomes, the following policy recommendations are made:

- Stakeholder engagement should be prioritized in Central and West Asia, the Advanced Economies, and Southeast Asia. Efforts should be made using formal and informal consultation mechanisms.
- Integrity can be improved in Central and West Asia and Southeast Asia by promoting legal and institutional frameworks to make decision-makers and stakeholders more accountable.
- Trade-offs involved in developing and managing water resources need more attention in Central and West Asia, South Asia, East Asia, and the Pacific. Public debate on the risks and costs associated with "too much, too little, or too polluted" water is needed to face the current uncertainties.
- Monitoring and evaluation should be improved in all regions with adequate institutional coordination. Promoting governance arrangements and helping water agencies across government levels raise the necessary revenues to meet their mandates should be prioritized, especially in the Pacific.

Water Security in the Advanced Economies in Asia

Water security in the Advanced Economies is overall good, but none of these economies are in the "the model stage" category. Rural household water security (KD1) and urban water security (KD3) are very high. Water-related disaster security (KD5) is also high despite regular occurring floods and droughts. There is room for improvement in economic water security (KD2) and environmental water security (KD4). East Asia's economic water security (KD2) score (particularly the PRC's) is higher than the scores in all the Advanced Economies. Priority actions for governance improvement in the Advanced Economies are policy coherence, monitoring and evaluation, and stakeholder engagement.

Finance. The financing needs to achieve water security are enormous. ADB's Strategy 2030 estimates the investment needs for water and sanitation to be on average $53 billion per year up to 2030, of which about one-third will be needed from the private sector. The OECD has analyzed the investment needs for water supply and sanitation (KD1 and KD3), irrigation infrastructure (part of KD2), and flood protection (part of KD5). The following policy recommendations—which seem even more pressing in the context of the COVID-19 pandemic that affects utilities' revenues and the fiscal capacity of most countries in the region—summarize priorities toward the achievement of water security:

- **Make the best use of available assets and financial resources.** Improving the operational efficiency and effectiveness of infrastructure and service providers or optimizing storage and conveyance systems can postpone new investments by extending the operational life of existing assets and providing better services.
- **Minimize future investment needs through policies promoting sustainable water resources management, water and wastewater infrastructure, policy coherence, cost-effective expenditure programs built on robust planning, and setting of priorities.** Innovation can contribute to minimizing costs and optimizing investment and should be exploited in line with adaptive capacities. Applying nature-based engineering might reduce investment and maintenance costs.
- **Harness additional sources of finance.** The huge investment deficit in the region requires leveraging financial resources from diverse potential sources. Governments need to increase contributions from polluters, users, and beneficiaries. They also need to increase reliance on domestic funds and attract private investments. Transitioning from concessional finance to crowding in commercial capital will be crucial.

At the time of writing, the full consequences of the COVID-19 pandemic are yet to be understood. In the short term, the health crisis emphasizes the urgent need to secure access to safe water for households, health-care facilities, schools, and public places. Secure and continuous water supply is essential for effective hygiene practices such as handwashing with soap, one of the most effective infection prevention and control measures for COVID-19 and other diseases. At the same time, the unfolding economic crisis is putting additional constraints on public finance, households' income, and utilities' revenues. These developments confirm the relevance of the three policy recommendations mentioned above and call for a more detailed assessment of implementation.

ADB Water Sector Framework and Focal Areas for Delivering ADB Support

The vision of ADB's water activities as formulated in the Strategic Directions for ADB Water 2030 is to achieve a water-secure and resilient Asia and the Pacific based on five integrating principles: (i) building resilience and adaptive capacity, (ii) promoting inclusiveness, (iii) embracing sustainability, (iv) improving governance, and (v) fostering innovation. Based on these principles, the following focal areas for delivering ADB support were formulated:

- **Water as a sustainable resource.** Policy initiatives will continue to promote regulatory and incentive mechanisms for improved governance and sustainable management of surface- and groundwater resources with the ultimate goal to support economic growth, restore aquatic ecosystems, and improve livelihoods.

- **Investment in universal water access.** Investing in universal and safe water services will be crucial, including access to reliable water supply and sanitation and sustainable wastewater management, appropriate to local conditions, through sewered and non-sewered, and centralized and decentralized systems. ADB will continue to support investments to create an enabling environment for private sector involvement and promote the circular economy approach by viewing waste as a resource.

- **Productive water in agriculture and energy.** ADB's irrigation investments will support diversified and higher value agriculture and more efficiency in water use to accelerate the contribution to rural revitalization and climate adaptation. Projects will be better integrated into a value chain approach and increasingly seek to ensure compatibility between energy and water resource planning.

- **Reduced water-related risk.** ADB's disaster risk reduction intervention will be integrated with other development programs, including livable cities and food security, by demonstrating risk-sensitive land use management approaches and nature-based solutions while integrating structural and nonstructural measures. This focal area includes developing clear strategies for building resilience to recurrent droughts and floods.

PART I
Introduction

AWDO's Historical Trajectories and Objectives

The Asian Water Development Outlook (AWDO) is a flagship publication of the Asian Development Bank (ADB) and the Asia-Pacific Water Forum to highlight critical water management issues for ADB members in Asia and the Pacific. AWDO is a living document that reflects evolving dynamics to track the region's water security. ADB published 2007, 2013, and 2016 AWDO editions, each building on the previous one to provide economy snapshots of Asia and the Pacific's water security status. The inaugural 2007 edition responded to the need to address water security, with a broader perspective than traditional sector-focused approaches. The 2013 edition introduced a multidimensional framework to assess water security, developing a rigorous quantitative methodology and transforming AWDO into an analytical tool. The 2016 edition refined the framework to make the analysis more robust and build greater confidence in assessing water security.

Throughout the years, AWDO has expanded progressively in its scope and approach. Building on the past editions, AWDO 2020 has become a policy-into-practice tool. First, it provides an up-to-date overview of the region's water security, showing progress and enabling methodological comparison over time. Second, it allows spatial comparison across different scales, including ADB members[1] and regions, by unpacking and disseminating scientific findings for larger audiences of policy makers, donor organizations, and private investors. Last, by generating targeted policy recommendations and testing AWDO methodology in ADB's selected developing members, AWDO

2020 will inform decision-making and public–private investments toward achieving water security in the region.

AWDO 2020 incorporates several elements of novelty. The breadth and depth of the analysis have expanded with methodological advancements, the greater granularity of policy recommendations across finance and governance aspects, and the introduction of three case country studies: Thailand, India's state Karnataka, and Timor-Leste. Ultimately, AWDO 2020 provides a stronger alignment with the Sustainable Development Goals (SDGs), as reflected in ADB's Strategy 2030 and the Strategic Directions for ADB Water 2030: Water-Secure and Resilient Asia Pacific.[2]

Development Challenges to Achieve a Water-Secure and Resilient Asia and the Pacific by 2030

Water underpins social and economic development. Managing water resources is becoming increasingly critical in Asia and the Pacific, with its fast-paced economic development, population growth, and changing climate. The urbanization trend will continue, resulting in an estimated 2.5 billion people, or 55% of the population, living in urban areas by 2030.[3] At the same time, although modern commercial agriculture is expanding, smallholder agriculture remains an integral part of rural economies in many countries, employing over 40% of people in South Asia and receiving on average 80% of Asia's freshwater. Poverty reduction

---

[1] ADB members as mentioned in the remaining text of this document refers to only the ADB members in the Asia and Pacific region.

[2] ADB. Forthcoming. *Strategic Directions for ADB Water 2030: Water-Secure and Resilient Asia Pacific.* Manila.

[3] United Nations Department of Economic and Social Affairs, Population Division. 2019. *World Urbanization Prospects: The 2018 Revision.* New York.

in Asia and the Pacific has been a successful story, dropping "from 53% in 1990 to about 9% of the total population in 2013."[4]

Despite this remarkable achievement, 326 million people still live under the poverty threshold of $1.90/day. The region is home to a total of 563 million urban slum dwellers, challenging the municipalities' ability to provide basic services.[5] About 300 million people in the region still have no access to safely managed or basic services of drinking water, and 1.2 billion lack adequate sanitation. Poor access to water and sanitation disproportionally impacts vulnerable groups, who are particularly susceptible to external economic shocks, e.g., rise in food prices or infectious diseases such as the coronavirus disease 2019 (COVID-19). Poverty itself is the single biggest determinant of health, with poor people consistently suffering poorer health and shorter life expectancies.[6]

These issues raise equity and human rights considerations. Traditional gender roles are largely associated with women's responsibility to fetch water, especially in rural households. Globally, it is estimated that 75% of the burden of collecting water falls on women. Although gender-disaggregated data are lacking regionally, Mongolia is the only country in Asia and the Pacific where men spend more time collecting water than women. Lack of appropriate, safe sanitation facilities creates more negative health, safety, and psychosocial stress impacts for women, further pronounced during menstruation. These factors lead to increased work absenteeism, school dropouts, and increased risks of assault in case of open defecation.[7]

Poor sanitation services are also correlated with negative health impacts such as diarrhea and malnutrition (Afghanistan and Pakistan). These effects are further exacerbated by tropical climate and reduced accessibility of health services (Pacific island countries). Therefore, improving access to adequate water supply and sanitation services to reach the last mile remains a development priority not only to reduce poverty but also to promote inclusiveness.

"Asia and the Pacific is the most disaster-affected region in the world, home to more than 40% of disasters and 84% of people affected."[5] Major delta cities along the coastlines are becoming increasingly vulnerable to climate change risks and disasters—flood events, sea level rise, and droughts—threatening lives, livelihoods, and public health and compounded with huge economic losses.

Poor water quality and increasing abstraction of water are putting pressure on the environment. Water quality in Asia has deteriorated significantly, with pollution increasing in 50% of major rivers during 1990–2010, salinity increasing by more than one-third, and 80% of wastewater being discharged into waterways without adequate treatment.[8] The rapid depletion of groundwater aquifers has led to inequities in water access, land subsidence in some major cities, and an increase in saltwater intrusion into coastal areas.

[4]   ADB. 2018. *Strategy 2030: Achieving a Prosperous, Inclusive, Resilient, and Sustainable Asia and the Pacific.* Manila. p. 3.

[5]   ADB. 2018. *Boosting Strategy 2030: Making Cities More Livable.* Manila.

[6]   World Health Organization. 2003. *Social Determinants of Health: The Solid Facts.* Copenhagen.

[7]   United Nations Educational, Scientific and Cultural Organization (UNESCO) World Water Assessment Programme. 2019. *The United Nations World Water Development Report 2019: Leaving No One Behind.* Paris.

[8]   United Nations Environment Programme. 2016. *A Snapshot of the World's Water Quality: Towards a Global Assessment.* Nairobi.

Many ADB members in the region face governance challenges and weak institutional capacity. Although governance and institutional performance have gradually improved since 2010, there remains a significant gap between economies in the Asia and Pacific and those of the Organisation for Economic Cooperation and Development (OECD), as reflected in the Worldwide Governance Indicators.[9] Poor governance is a particularly challenging issue in service delivery and infrastructure provision, and service reliability and affordability.

Future Water Security Risks

This document describes the current water security of the regional ADB members. All members take action to improve their water security through infrastructure investments and more efficient management of their resources. However, external conditions might change, posing a risk to water security. Socioeconomic development might lead to increased water demand and pollution, reducing water security, as much as it creates higher expectations for and more resources to deliver on water security. Climate change is changing water availability and variability. Political contexts might also change, resulting in higher or lower prioritization of water security investment and reduced attention governments give to the environmental component of water security. Then the COVID-19 pandemic poses a range of direct and indirect risks to water security. These uncertainties are not taken into account in the scores of water security in this report. Although not reflected directly in the scores, future water security risks are addressed in this document in the various water security components.

ADB's Vision on Water Security and AWDO 2020

Defining Water Security toward Achieving Sustainable Development Goal 6 by 2030

The concept of water security has developed over time, from a general vision to a goal to be achieved with integrated water resources management (IWRM). It is still contested and evolving. The Global Water Partnership introduced the concept of IWRM[10] as "a process which promotes the coordinated development and management of water, land, and related resources in order to maximise the resultant economic and social welfare in an equitable manner without compromising the sustainability of vital ecosystems."[11] IWRM is applied at a basin scale to include water management's upstream and downstream aspects, encompassing water management's human, sociopolitical, and ecological dimensions. Tying all these elements together allows a better balance of water management as a service and a resource. If IWRM is the process or journey, water security is one of the important outcomes.[12]

One of the most widely cited and used definitions of water security is this: "the availability of an acceptable quantity and quality of water for health, livelihoods, ecosystems and production, coupled with an acceptable level of water-related risks to people, environments and economies."[13]

9   ADB. 2019. *Strategy 2030 Operational Plan for Priority 6: Strengthening Governance and Institutional Capacity, 2019–2024*. Manila.

10   Global Water Partnership Technical Advisory Committee (TAC). 2000. Integrated Water Resources Management. *TAC Background Paper*. No. 4. Stockholm.

11   Beek, E. van and W. Lincklaen Arriens. 2014. Water Security: Putting the Concept into Practice. *TEC Background Paper*. No. 20. Stockholm: Global Water Partnership Technical Committee (TEC). p. 23.

12   Grey, D. 2019. Reflections on Water Security and Humanity. In Dadson, S. J. et al., eds. *Water Science, Policy, and Management: A Global Challenge*. University of Oxford, School of Geography and the Environment.

13   Grey, D. and C. W. Sadoff. 2007. Sink or Swim? Water Security for Growth and Development. *Water Policy*. 9 (6). pp. 545–571.

This broad definition has two essential elements: (i) the multidimensional nature of water security across different water uses; and (ii) a risk-based approach to inform how societies cope with water-related risks including floods, droughts, and contamination. Complementary to this definition is its slightly revised version: "Water security is a tolerable level of water-related risk to society."[14] Substituting the adjective "acceptable" with "tolerable," this definition emphasizes water security's community-specific social, economic, and cultural values. Water risks usually become less tolerable with increasing levels of economic growth and wealth. Thus, perspectives may change depending on socioeconomic conditions over time and across geographies, making the concept of water security dynamic.

These definitions reflect the changing global policy dynamics on adopting the 2030 Agenda for Sustainable Development by the United Nations in 2015. If Millennium Development Goal 7 (target 7.3) focused on access to safe drinking water supply and basic sanitation, the Sustainable Development Goals (SDGs) represented a paradigm shift. SDG 6 (clean water and sanitation) looks beyond drinking water supply and sanitation (targets 6.1 and 6.2) to encompass aspects of water quality and wastewater (target 6.3), water use and efficiency (target 6.4), IWRM (target 6.5), ecosystems (target 6.6), and enabling environment (targets 6.a and 6.b). This changed mindset reflects an acknowledgment of the complexities of water resources management and the urge to revisit existing paradigms with integrated approaches by bringing together different sectors and stakeholders. To address current challenges, "a rapid change of the economics, engineering and management frameworks that guided water policy and investments in the past"[15] is needed, bringing resilience, governance, and innovative finance into the management equation.

AWDO Definition of Water Security across Five Key Dimensions

Against this backdrop, ADB developed a framework with five interdependent key dimensions (KDs) to quantify water security in Asia and the Pacific. These KDs are illustrated in Figure 1 and described more in detail in the next section. In developing the initial analytical framework for AWDO since 2013, ADB formulated a shared vision of water security with the following logic: societies can enjoy water security when they successfully manage their water resources and services to (i) satisfy rural household water and sanitation needs in all communities; (ii) support productive economies in agriculture, industry, and energy; (iii) develop vibrant, livable cities and towns; (iv) restore healthy rivers and ecosystems; and (v) build resilient communities that can cope with water-related extreme events.

Combining these elements results in the AWDO definition of water security: the availability of an adequate quantity and quality of water to ensure safe, affordable, equitable, and inclusive water supply and sanitation, together with sustainable livelihoods and healthy ecosystems and manageable water-related risks. Operationalizing water security will help foster resilient rural-urban economies in Asia and the Pacific.

---

[14]   Grey, D. et al. 2013. Water Security in One Blue Planet: Twenty-First Century Policy Challenges for Science. *Philosophical Transactions of the Royal Society.* A 371.

[15]   Sadoff, C. W., E. Borgomeo, and S. Uhlenbrook. 2020. Rethinking Water for SDG 6. *Nature Sustainability.* 3. pp. 346–347.

Figure 1: Water Security Framework of Five Interdependent Key Dimensions

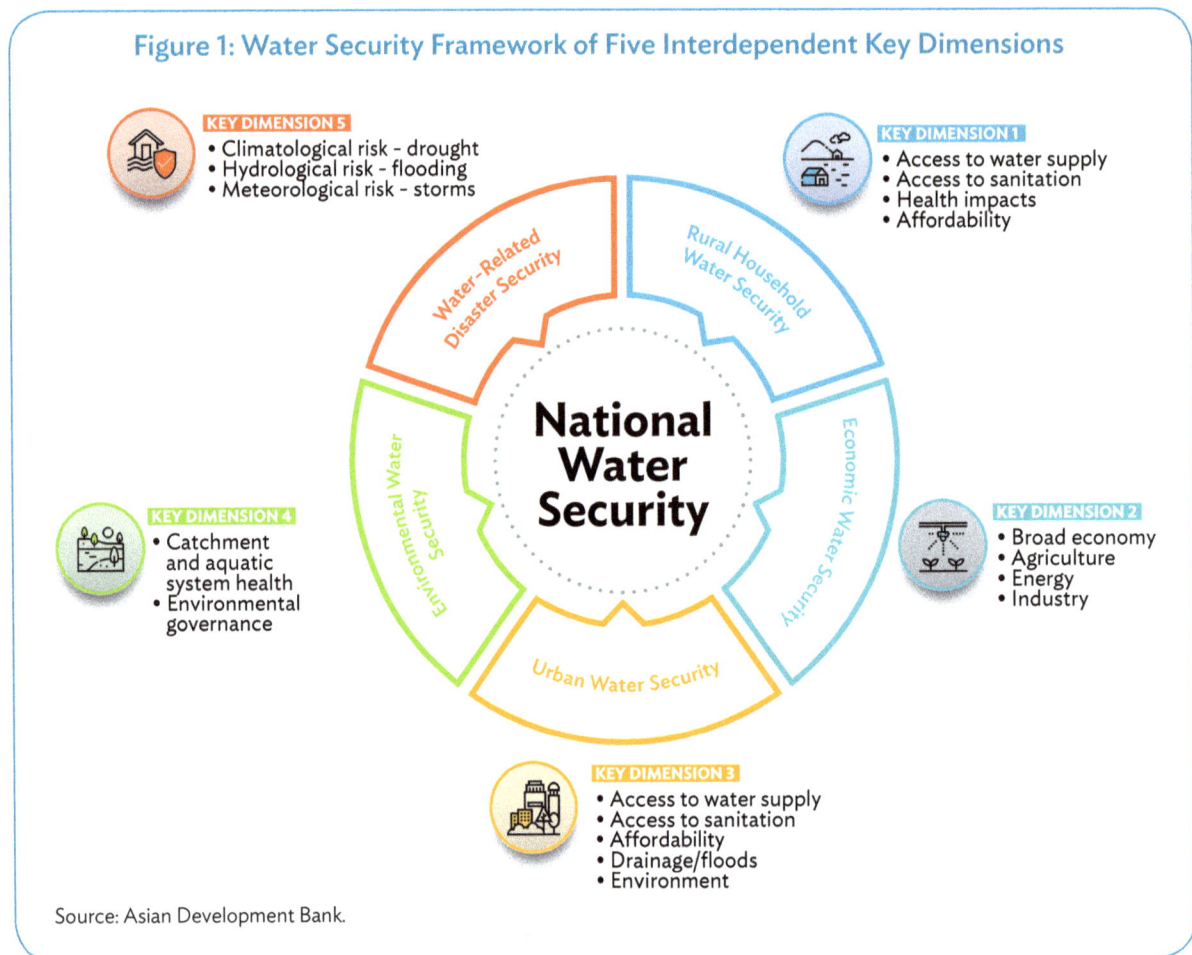

KEY DIMENSION 5
- Climatological risk - drought
- Hydrological risk - flooding
- Meteorological risk - storms

KEY DIMENSION 1
- Access to water supply
- Access to sanitation
- Health impacts
- Affordability

KEY DIMENSION 4
- Catchment and aquatic system health
- Environmental governance

KEY DIMENSION 2
- Broad economy
- Agriculture
- Energy
- Industry

KEY DIMENSION 3
- Access to water supply
- Access to sanitation
- Affordability
- Drainage/floods
- Environment

Water-Related Disaster Security

Rural Household Water Security

Environmental Water Security

Economic Water Security

Urban Water Security

National Water Security

Source: Asian Development Bank.

AWDO Framework and National Water Security Index

Each country's overall national water security (NWS) is assessed as the composite result of the five KDs (KD1–KD5). AWDO measures water security by quantifying the five KDs in terms of clear and measurable indicators (Table 1).

Table 1 illustrates that the five KDs of water security are related, interdependent, and should not be treated in isolation. Measuring water security by aggregating indicators in these KDs recognizes their independencies. Increasing water security in one dimension may simultaneously increase or decrease security in another dimension and affect overall NWS. Given the interdependence of the factors determining water security in each dimension,

increases in water security will be achieved by governments that break the traditional sector silos to find ways of managing the linkages, synergies, and trade-offs among the dimensions with holistic and integrated approaches.

National Water Security Stages

The five KDs form the National Water Security Index (NWSI). Appendix 1 provides the scores of the five KDs for all 49 regional ADB members. The maximum score for each KD is 20. The maximum NWS score—the sum of the KDs—is 100. The five stages of water security assessment are summarized in Table 2. At NWSI Stage 1, the water situation is nascent, and there is a large gap between the current state and the acceptable level of water

Table 1: AWDO Framework for Assessing National Water Security

Logos	KD	Index	Measurement	Composition
		National Water Security Index (NWSI)	The availability of adequate water to ensure safe and affordable water supply, inclusive sanitation for all, people's livelihoods and healthy ecosystems, with reduced water-related risks toward supporting sustainable resilient rural–urban economies	Total of the five dimensions of water security
	Key dimension 1 (KD1)	Rural household water security	The provision of sufficient, safe, physically accessible,[a] and affordable water and sanitation services for health and livelihoods, coupled with an acceptable level of water-related risk, in rural households	Access to water supply Access to sanitation Health impacts affordability
	Key dimension 2 (KD2)	Economic water security	The assurance of adequate water to sustainably satisfy a country's economic growth and avoid economic losses due to water-induced disasters	Broad economy agriculture Energy Industry
	Key dimension 3 (KD3)	Urban water security	The extent Asian Development Bank members provide safely managed and affordable water and sanitation services for their urban communities to sustainably achieve desired outcomes	Access to water supply Access to sanitation Affordability Drainage (flooding) Environment (water quality)
	Key dimension 4 (KD4)	Environmental water security	The health of rivers, wetlands, and groundwater systems and measured progress on restoring aquatic ecosystems to health on a national and regional scale	Catchment and Aquatic System Condition Index Environmental Governance Index
	Key dimension 5 (KD5)	Water-related disaster security	A nation's recent exposure to water-related disasters, their vulnerability to those disasters, and their capacity to resist and bounce back	Climate risk (drought) Hydrological risk (flood) Meteorological risk (storm)

[a] Accessibility distinguishes the various steps in the service ladder as defined in the Joint Monitoring Programme.
Source: Asian Development Bank.

security. At NWSI Stage 5, the country may be considered a model for its management of water services and resources, and as water secure as possible under current circumstances.

AWDO 2020 stages are the same as in AWDO 2016 and AWDO 2013. The lower banding has been changed somewhat to account for the methodology changes applied to determine the KD scores. The defined bands do not have a hard scientific substantiation but are based on expert judgment of the AWDO 2020 team and composed of the bandings of the individual KDs. The bandings applied in the KDs are described in Appendixes 3–7.

Table 2: National Water Security Stages

NWSI	NWS Score	NWS Stage	Description
5	96 and above	Model	All people have access to safe, affordable, and reliable drinking water and sanitation facilities. Economic activities are not constrained by water availability. Environmental governance is good, and pressure on aquatic ecosystems is limited. Water-related risks are acceptable and relatively easy to deal with.
4	78–96	Effective	Nearly all people have access to affordable safe drinking water and sanitation facilities. Economic water security is high. Environmental governance is generally acceptable, and attention is given to ecological restoration. There are systematic commitments to reduce disaster risk.
3	60–78	Capable	Access to safe drinking water and sanitation facilities is improving. Economic water security is moderate. Environmental governance is moderate, with clear pressure on the ecosystem. There are some institutional commitments to reduce disaster risk.
2	42–60	Engaged	A significant majority of rural and urban households have access to basic water supply but less to sanitation. Economic water security is low. Environmental governance is moderate, with severe pressures on aquatic ecosystems. Progress in achieving disaster risk security is low.
1	0–42	Nascent	A low proportion of rural and urban households have access to basic water supply and sanitation. Economic water security is low. Environmental governance is poor, with significant pressures on the aquatic ecosystems. Hardly any attention is given to disaster risk reduction.

NWS = national water security, NWSI = National Water Security Index.
Source: Asian Development Bank.

Scoring Approach in AWDO 2020

The country's performance in the KDs is expressed in scores (see the appendixes for the summary).

More details are given in the methodology and data report of AWDO 2020.[16] Each KD is scored on a scale of 1–20. The NWS score (1–100) is the sum of the five KD scores.

---

[16]   ADB. Forthcoming. AWDO 2020: Description of Methodology and Data.

Methodology Changes between AWDO 2020 and AWDO 2016

Based on the experience with AWDO 2016, ADB introduced several changes in the methodology framework of AWDO 2020. These changes are listed in Appendixes 3–7 and described in detail in AWDO 2020's methodology report:

- KD1 is redefined as **rural** household water security. KD1 in AWDO 2016 included urban areas. This redefinition addresses the "double counting" in KD3 in the previous versions.
- **KD1** and **KD3** provide a more granular indicator for the water supply index considering the **service ladder** distinguishing basic and safely managed services.
- KD5 is based on a risk approach and includes recent **hazard** impacts. KD5 in AWDO 2016 addressed resilience only.
- **Future risk** is included in the narratives of all KDs.

In addition to these methodology changes, two important elements of novelty in AWDO 2020 are the

- analysis of governance and finance as crosscutting themes; and
- AWDO framework application to three country case studies: Thailand, India's Karnataka, and Timor-Leste.

Other changes are (i) Timor-Leste has moved from the Pacific to Southeast Asia, and (ii) Niue has been included again in the analysis.

Due to these changes in methodologies and novelty elements, the results as published in the AWDO 2013 and AWDO 2016 reports are not directly comparable with the results of AWDO 2020. To investigate the impacts of the methodology changes, ADB applied the new 2020 methodology on the data used for AWDO 2013 and AWDO 2016, comparing AWDO 2013 and AWDO 2016 results using the old and new methodologies. This comparison is described in the methodology report. The analysis shows that the methodology update has some impacts but that these did not change the conclusions that were drawn in the AWDO 2013 and AWDO 2016 reports. The 2013 and 2016 results included in AWDO 2020 are all based on the AWDO 2020 methodology, making it possible to assess the progress from 2013 to 2016 to 2020.

Harvesting rice in Bojong Village near Yogyakarta, Indonesia

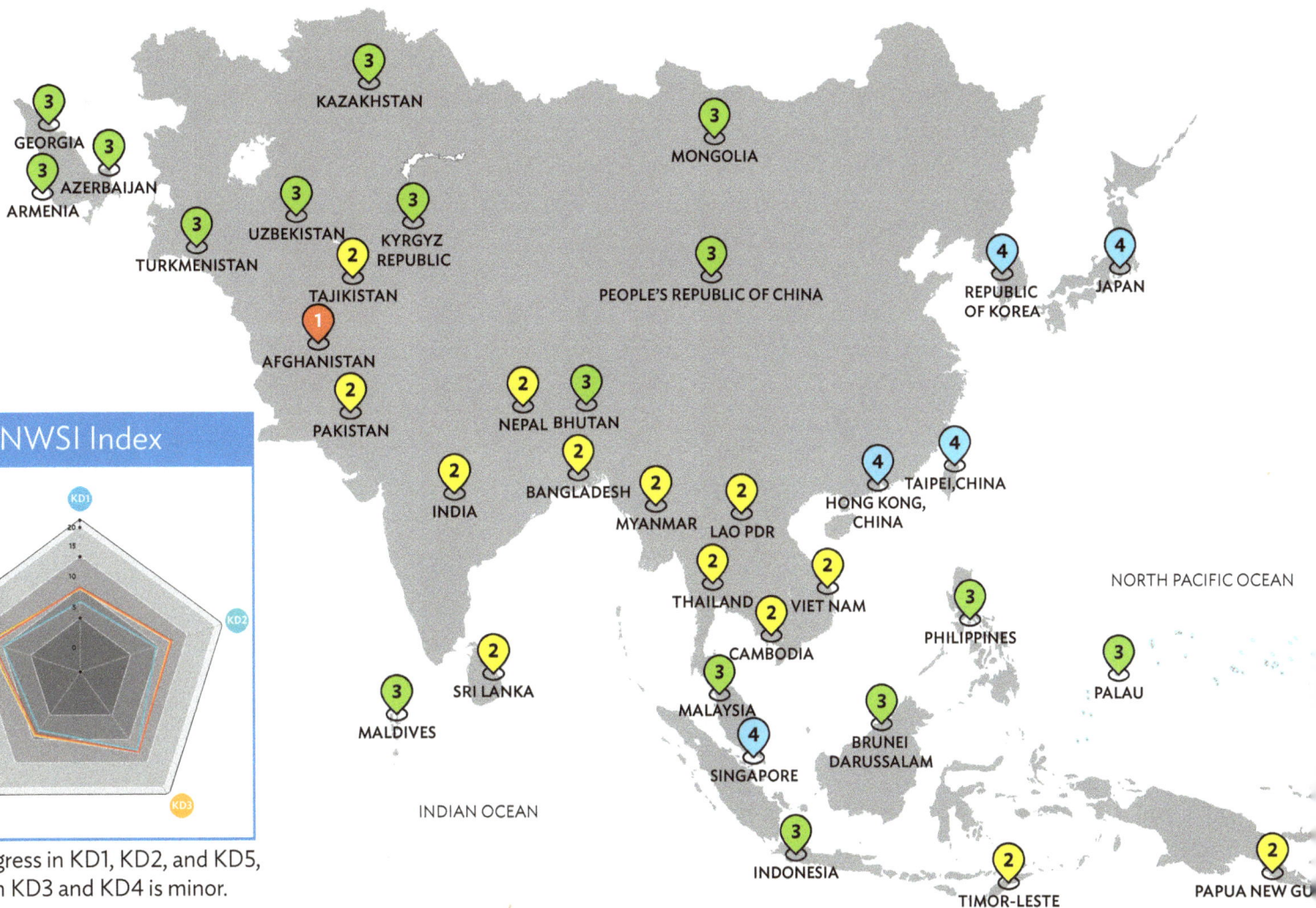

KAZAKHSTAN **3**

MONGOLIA **3**

GEORGIA **3**
AZERBAIJAN **3**
ARMENIA **3**

TURKMENISTAN **3**
UZBEKISTAN **3**
KYRGYZ REPUBLIC **3**
TAJIKISTAN **2**
AFGHANISTAN **1**

PEOPLE'S REPUBLIC OF CHINA **3**

REPUBLIC OF KOREA **4**
JAPAN **4**

PAKISTAN **2**
NEPAL **2**
BHUTAN **3**

INDIA **2**
BANGLADESH **2**
MYANMAR **2**
LAO PDR **2**
HONG KONG, CHINA **4**
TAIPEI,CHINA **4**

THAILAND **2**
VIET NAM **2**
CAMBODIA **2**
PHILIPPINES **3**

NORTH PACIFIC OCEAN

SRI LANKA **2**
MALDIVES **3**
MALAYSIA **3**
SINGAPORE **4**
BRUNEI DARUSSALAM **3**
PALAU **3**

INDIAN OCEAN

INDONESIA **3**
TIMOR-LESTE **2**
PAPUA NEW GU **2**

NWSI Index

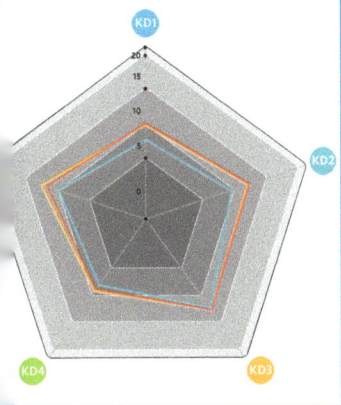

KD1
20
15
10
0
KD2
KD4
KD3

progress in KD1, KD2, and KD5,
ss in KD3 and KD4 is minor.

NS – AWDO 2013 to 2020

KD5
PRC (+6)
Kyrgyz Republic (+5)
Niue (+4)
Taipei,China (+4)
Tajikistan (+4)

KD4
Marshall Islands (+8)
Maldives (+5)
Philippines (+4)
Taipei,China (+4)
Tuvalu (+4)

(out of a max scale of 20)

TOP PERFORMERS ON NWS – AWDO 2013 to 2020

PRC (+16)
Kyrgyz Republic (+10)
Maldives (+10)
Palau (+10)
Philippines (+9)
Tajikistan (+9)
Taipei,China (8)
Kazakhstan (+7)
Bangladesh (+6)
Cook Islands (+6)
Lao PDR (+6)
Marshall Islands (+6)
Mongolia (+6)
Niue (+6)
Samoa (+6)
Tuvalu (+6)
Uzbekistan (+6)
Viet Nam (+6)

(out of a max scale of 100)

1 Nascent   **2** Engaged   Cap

Water supply facilities installed as part of ADB's Tonle Sap Rural Water
Supply and Sanitation Project in Kampong Chhnang Province, Cambodia

PART II
Water Security across Five Key Dimensions

AWDO 2020 builds on the work of AWDO 2013 and AWDO 2016 in analyzing Asia and the Pacific's water security. It has refined the underlying methodologies to quantify the five KDs and used the most recent data.[17] The following sections present the water security status of 49 ADB members in Asia and the Pacific, discussing the results and policy recommendations on increasing water security. The actual data (scores) by economy are in Appendixes 1–7. Appendix 1 provides an overview of the five KDs and the NWSI by economy. Appendix 2 presents the results by region, including regional average scores. Appendixes 3–7 show the scores for KDs 1–5. Appendix 8 provides the databases used to derive the scores.

The actual data years of the parameters considered in AWDO differ. For some parameters, recent data were used; for others, only data of earlier years were available. As a rule of thumb, one might consider that AWDO 2020 describes the situation in 2018, AWDO 2016 in 2014, and AWDO 2013 (due to a long publishing process) in 2009. Thus, comparing AWDO 2020 with AWDO 2016 and AWDO 2013 shows the progress made across 4–5-year span.

In describing the performance of water security KDs, ADB analyzes the relationship between water security and the gross national income (GNI) per capita. Does a country have to be rich to achieve higher water security? Or does higher water security contribute to more welfare? Graphs in the following section show the relationships between the two.

National Water Security Index

National Water Security Index: All Key Dimensions

The five KDs form the NWSI. Appendix 1 provides the scores of the five KDs for all 49 ADB members. The maximum score for each KD is 20. The maximum score for NWSI, the sum of the five KDs, is 100. The range in scores is enormous—from 39.5 (Afghanistan) to 89.1 (New Zealand)—as illustrated in Figure 2, in which ADB members are sorted based on their NWSI. The range is rather continuous. No clear groups can be identified. A low score does not mean that no progress is made (Box 1), also illustrated in the case of Afghanistan in Box 2 and of India in Box 3.

Box 1: Scores and Progress Are Different Things

A low position on the score list does not mean that no progress is made in water security. For example, Afghanistan, although at the bottom of the list, has made significant progress during 2013–2020 in rural and urban water security (KD1 and KD3). At the same time, New Zealand, at the top of the list, saw their urban water security decrease over that same period. Another top country, Australia, saw their overall national water security decrease between 2016 and 2020.

KD = key dimension, KD1 = rural household water security, KD3 = urban water security.
Source: Asian Development Bank.

---

[17] The methodology and data used for AWDO 2020 are described in the forthcoming ADB report.

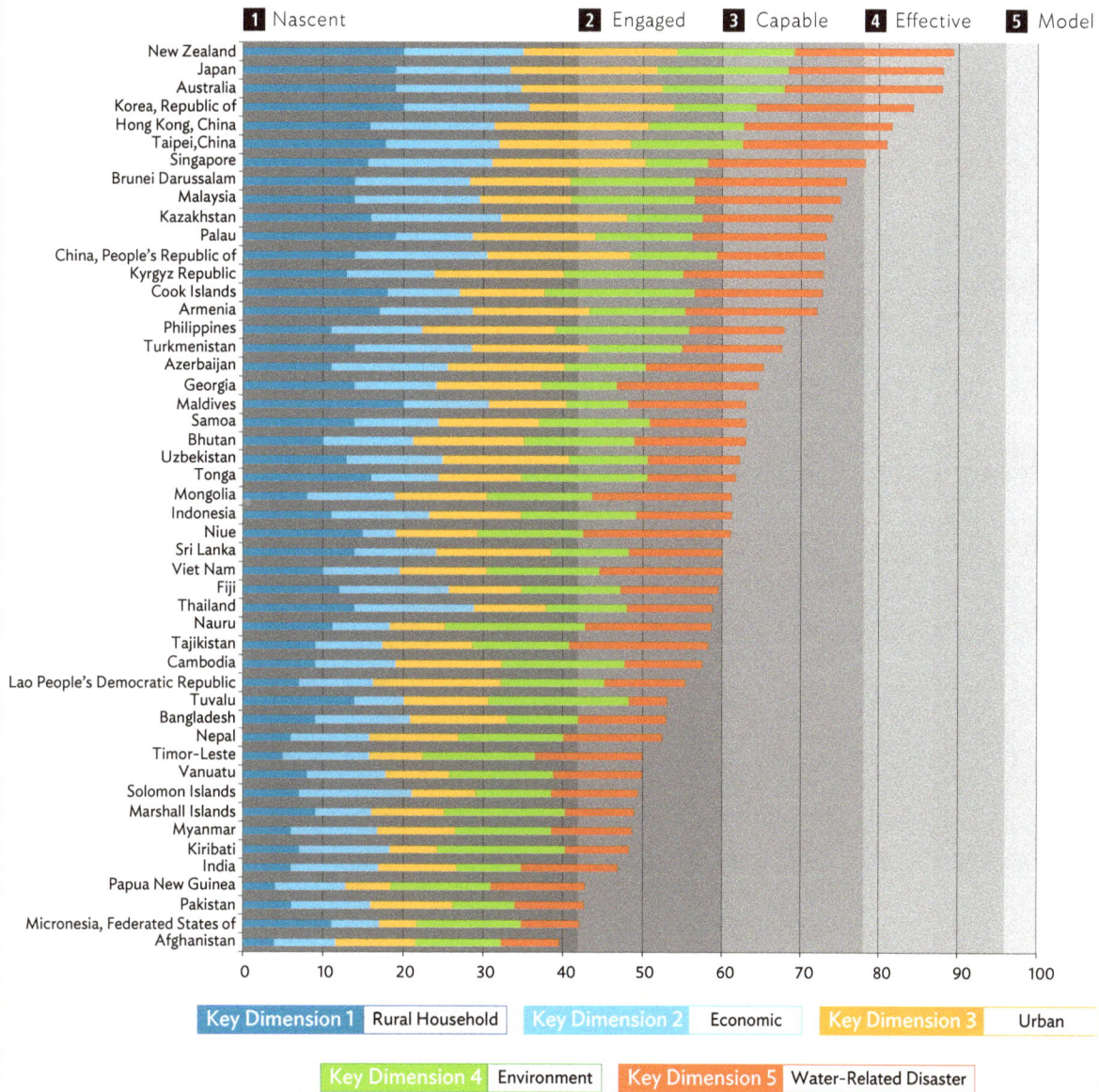

Figure 2: National Water Security Scores

Source: Asian Development Bank.

Figure 3 shows the scores of the five KDs in Asia and the Pacific for 2013, 2016, and 2020. These scores are population-weighted averages in the region, excluding the Advanced Economies. Good progress is seen between 2013 and 2020, particularly for KD1, KD2, and KD5.

National Water Security Stages

A similar positive development is seen if the NWS is expressed in terms of the five NWS stages, as defined in Table 2: nascent, engaged, capable, effective, and model.

Figure 3: Average Scores of Key Dimensions in Asia and the Pacific, 2013-2020

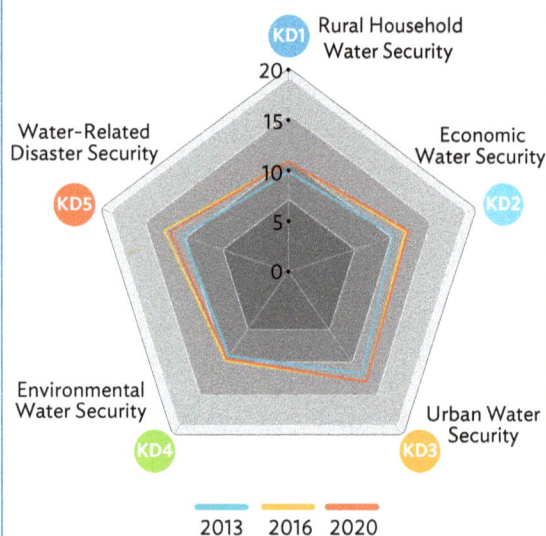

KD = key dimension.
Source: Asian Development Bank.

The left part of Figure 4 illustrates that 22 of 49 ADB members are still in the nascent and engaged stages in AWDO 2020. The development over time is promising, showing that ADB members move from nascent to engaged, and from engaged to capable and effective. No country has reached the model level yet, not even the Advanced Economies.

The right side of Figure 4 shows the population of ADB members in these development levels.

National Water Security across the Regions

Figure 5 presents the population-weighted averages of NWS by region.[18] The regional average for 2020 does not include the Advanced Economies. The population-weighted score of the Advanced Economies in AWDO 2020 is 86.5, the Pacific only 45.4, and South Asia 47.7. East Asia shows a promising score of 72.8, mainly due to the combined high scores of Taipei,China (80.8) and the People's Republic of China (PRC) (72.7). The figure shows that progress has been made in all regions between 2013 and 2020.

Figure 4: Number of ADB Members and People in Five National Water Security Stages

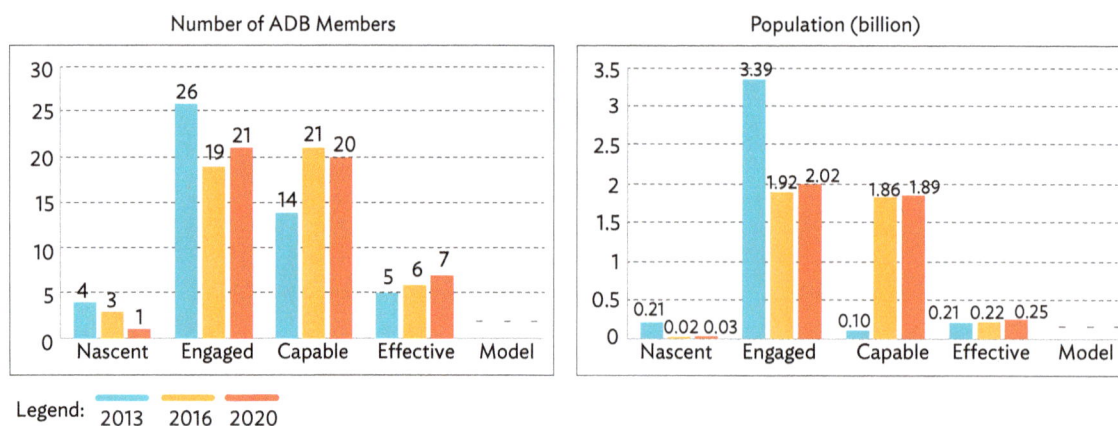

Legend: 2013 2016 2020

Source: Asian Development Bank.

---

[18]   See Appendix 1 for an overview of the ADB members in the six regions and an explanation of the population-weighting approach used in AWDO.

Figure 5: National Water Security by Region

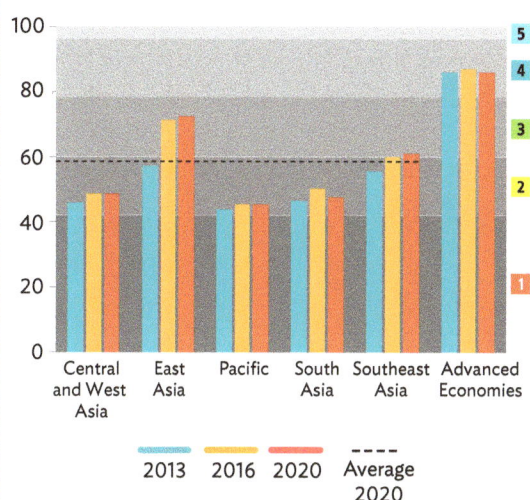

Source: Asian Development Bank.

Risks to Future National Water Security

Political will is needed toward improving governance to continue the progress in NWS, which involves maintaining assets and making necessary investments.

ADB members in the region have to be prepared to deal with potential risks. AWDO 2020 takes into account the multidimensional nature of risk, encompassing physical risks (e.g., water-related disasters, climate change, ecological and health risks); institutional and regulatory risks (e.g., change of policy and regulatory frameworks, shifts in markets, norms, and technologies); and financial risks (e.g., infrastructure and operation and maintenance costs), along with their associated negative socioeconomic impacts on people's assets and livelihoods. The main risk elements to be considered in water security are the impacts of the COVID-19 pandemic and the long-term effects of climate change.

The COVID-19 pandemic has emerged as a major short- and medium-term risk for water security, with direct risks to health, morbidity, and mortality and to the continuity and quality of water service provision due to human resources and supply chain interruptions. Indirect risks are the reduced financial resources for capital expansion and maintenance of infrastructure with secondary impacts on livelihoods and well-being, including increased poverty, malnutrition from reduced household income, and reduced educational achievement of children due to school closures, with possible long-term impacts on their adult lives. Another way to look at these risks is that improving water security (KD1 and KD3) will facilitate improved hygiene practices to mitigate the spread of COVID-19.

Climate change is expected to impact all water security KDs, with changing rainfall and runoff patterns impacting water availability to meet water demand, affecting rural household water security (KD1), economic water security (KD2), and urban water security (KD3). Aquatic ecosystems, and thus environmental water security (KD4), are predicted to face escalating negative impacts on many levels. More intense rainfall and typhoons, sea level rise, and more extreme droughts will affect water-related disaster security (KD5).

Addressing these risks might imply that trade-offs are needed between the KDs, e.g., more water storage to adapt to climate change will support KD2 but might negatively influence KD4. The necessary governance action to address these trade-offs is explained in the section on Improving Water Security and Key Dimension Performance by Good Governance. More specific risk elements across the five KDs are described under each of these KDs.

National Water Security and Gross National Income

The relation between the NWS of ADB members and their gross national income (GNI) per capita is shown in Figure 6. The GNI is plotted at a logarithmic scale. As expected, the figure shows a strong correlation between the two parameters. Particularly interesting are the ADB members that score positively in relation to the regression line and those that find themselves below the line. The Kyrgyz Republic performs very well considering its GNI and, to a lesser extent, so do Uzbekistan, the Philippines, and Armenia. On the other side of the spectrum, Papua New Guinea is clearly lagging, followed by Pakistan, India, and Maldives. Thailand also scores surprisingly low. As mentioned, a low score does not mean that no progress is made as illustrated for Afghanistan (Box 2) and India (Box 3).

Figure 6 is not population weighted. This means that Maldives, with a population of only 0.5 million, has the same influence on the regression line as the PRC (1.4 billion) and India (1.3 billion). The Pacific developing member countries (DMCs) are not included for this reason.

Box 2: Major Water Security Progress in Afghanistan

Despite major progress made in Afghanistan since 2013, the country remains at the nascent stage in the Asian Development Outlook (AWDO) 2020 mainly because of its fragile and conflict-affected situation over the last 4 decades, which has resulted in degraded infrastructure, limited human resources capacities, and a lack of institutional establishment. Progress has been achieved through strong partnerships between the Government of Afghanistan, community leaders, and international development partners. This progress includes rehabilitating and upgrading irrigation schemes, improving hydrometeorological and hydrogeological networks, implementing national dam safety guidelines and manuals, preparing water sector development strategies and master plans, developing new water regulatory law and several bylaws, and establishing the National Water Affairs Regulation Authority (the government body responsible for water resources management) in 2020. These accomplishments have already contributed to an increase in agricultural yield and farm household income. Crop productivity has increased by at least 20% for major crops in rehabilitated schemes. Further improvements in all five key dimensions (KDs) of water security are the government's overarching priorities.

Source: Asian Development Bank.

Box 3: Working toward Higher Water Security in India

Recognizing the importance of water security for the country's socioeconomic development, the Government of India along with state governments have embarked on several extensive investment programs in the water sector. In addition to several state government funded programs in the water sector, some of the notable central government funded programs include the Swachh Bharat Mission (SBM) (elimination of open defecation and improvement of solid waste management); Jal Jeevan Mission (JJM) (provision of functional tap connection to every rural household); Atal Mission for Rejuvenation and Urban Transformation (AMRUT); Pradhan Mantri Krishi Sinchayee Yojana (PMKSY); Smart Cities, National Hydrology Project (NHP); Dam Rehabilitation and Improvement Project (DRIP); National Mission for Clean Ganga (NMCG); Namami Gange Program; National River Conservation Plan (NRCP); Atal Bhujal Yojana (ABJY); MNREGS; and Disarmament, Demobilization, and Reintegration programs (following the Hyogo Framework), etc.

Results of these investment programs show the progress made in comparison to other countries:

- India's improvement in sanitation during 2017–2019 was three times greater than the next best country considered in the Asian Water Development Outlook (AWDO) 2020.

continued on next page

Box 3 continued

- Over 80% of rural Indians that needed to openly defecate in 2014 have access to basic sanitation in 2019.
- 100,000 house connections are realized each day with a target to cover more than 180 million house connections till 2024.
- PMKSY comprises of components including the creation of Major/Medium/Minor storage projects, Command Area Development Works, Watershed Development, as well as Per Drop More Crop (water efficiency works).
- In addition to PMKSY, the Government of India is also funding several national projects to improve water storage capacities in the country which are under construction/ implementation.
- Some notable water conservation programs implemented by state government includes: Jalyukt Shivar (Maharashtra), Mukhya Mantri Jal Swavalamban Abhiyan (Rajasthan), NEERU CHETTU (AP), Mission Kakatiya (Telangana), Sujalam Sufalam Yojana (Gujarat), IWRM and Artificial Recharge Structures Scheme (Karnataka).

The impact of all these programs along with several similar programs funded by state and central governments may not be fully captured in the data sets used for AWDO 2020 resulting in an underestimation of key dimensions scores but will certainly result in higher key dimension scores in the next AWDO.

Source: Asian Development Bank.

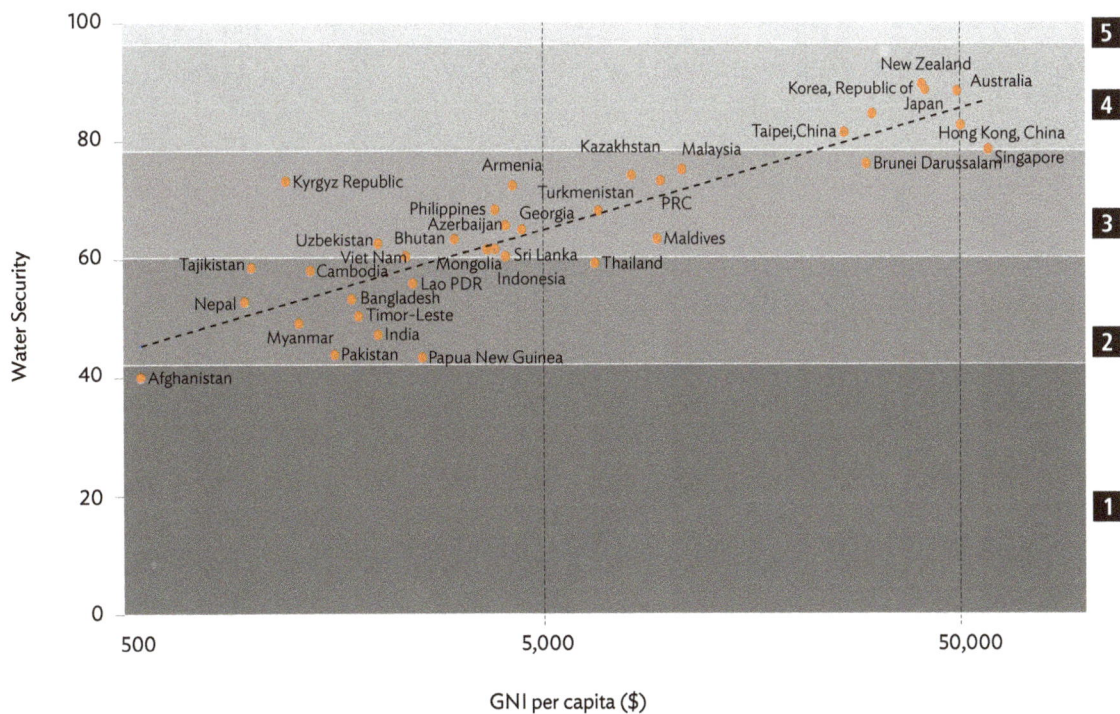

Figure 6: National Water Security and Gross National Income per Capita, 2020

GNI = gross national income, Lao PDR = Lao People's Democratic Republic, PRC = People's Republic of China.
Note: Small Pacific island states are not included.
Source: Asian Development Bank.

Community piped water system implemented as part of ADB's
Chittagong Hill Tracts Rural Development Project in Bangladesh

RURAL HOUSEHOLD WATER SECURITY

KD1 assesses the extent to which ADB members provide sufficient, safe, physically accessible, and affordable water and sanitation services for health and livelihoods, coupled with an acceptable level of water-related risk, in rural households.

Selected KD1-related projects

India
West Bengal Drinking Water Sector Project

The project will provide safe, sustainable, and inclusive drinking water services to about 1.65 million people in selected areas of West Bengal. It will introduce an innovative institutional framework and technology for smart water management to enable efficient service delivery.

People's Republic of China
Hunan Miluo River Disaster Risk Management and Comprehensive Environment Improvement Project

This integrated project in Pingjiang County includes the construction of a drinking water supply plant and supply of safe drinking water, separation of rainwater from sewers, establishment of concentrated rural wastewater treatment services, and upgrading of rural toilets.

Indicators included in KD1

- Access to water supply
- Access to sanitation
- Health impacts related to water
- Affordability

KD1 direct links to the SDGs

SDG 6

TARGET 6.1: By 2030, achieve universal and equitable access to safe and affordable drinking water for all.

TARGET 6.2: By 2030, achieve access to adequate and equitable sanitation and hygiene for all, and end open defecation, paying special attention to the needs of women and girls and those in vulnerable situations.

And indirectly contributes to:

- SDG 1 — No poverty
- SDG 3 — Good health

What do the numbers tell us about KD1?

KD1 Score

Legend: 2013 · 2016 · 2020 · Average 2020

Categories (x-axis): Central and West Asia, East Asia, Pacific, South Asia, Southeast Asia, Average 2020

Number of ADB Members

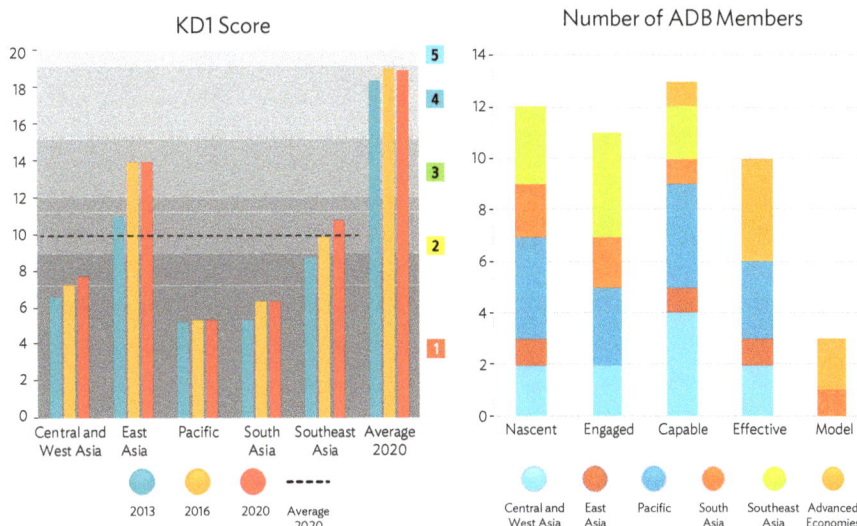

Categories (x-axis): Nascent, Engaged, Capable, Effective, Model

Legend: Central and West Asia · East Asia · Pacific · South Asia · Southeast Asia · Advanced Economies

All regions show progress from 2013 to 2020

12 ADB members are still in the nascent development stage, including India, meaning that 1.1 billion rural people are still facing rural household water insecurity

Performance is best in East Asia, followed by Southeast Asia, the Pacific and South Asia lag somewhat behind, with South Asia showing good progress

Top performers and challenged ADB members on KD1

Between 2013 and 2020 (scores up at scale of 20)

+5 points
Maldives

+4 points
Armenia • Kazakhstan

+2 points
Bhutan • Cambodia • People's Republic of China • Fiji • Kyrgyz Republic • Lao PDR • Palau • Philippines • Sri Lanka

Lower score in 2020

−1 point
Australia • Marshall Islands • Mongolia

Key Dimension 1: Rural Household Water Security

Introduction

In AWDO 2020, KD1 has been redefined from household water security in 2016 to rural household water security. Specifying rural areas meant that the KD needed a new definition.

KD1: Rural Household Water Security

Rural household water security is about providing sufficient, safe, physically accessible, and affordable water and sanitation services for health and livelihoods, coupled with an acceptable level of water-related risk, in rural households.

Typically, rural households are poorer and more disenfranchised than urban households. Urban households usually have higher disposable incomes and better access to service provision than rural households. Further, investing in water and sanitation for households is generally less attractive to funding organizations, as the return on investment is lower and indirect, compared with investing in water for economic uses like agriculture. It could be argued that rural households are the most vulnerable to water-related risks and, with lower service provision, constitute the least water-secure communities.

A key element emphasized by the Sustainable Development Goals (SDGs) is that no one is left behind. Rural households, along with those living in urban informal settlements (considered part of KD3), are the most vulnerable in the AWDO. Thus,

of all the KDs, KD1 is most closely aligned with SDG 6 (clean water and sanitation), particularly with targets 6.1 (safe drinking water) and 6.2 (safe sanitation and hygiene). Additionally, improvements in water, sanitation, and hygiene (WASH) often lead to improvements in both education (SDG 4) and gender equity (SDG 5). Investment in WASH has shown to generate a very high return on investment through health, education, and other benefits.[7]

Governments and donors must recognize that improving rural household water security has its own intrinsic value, which is inherently linked to a holistic definition of water security.

As the definition of KD1 changed, the method used for calculating the index score also changed. A risk framework with four indicators has been developed to create the 2020 KD1 score:

- Indicator 1 (access to water supply)—the percentage of rural people with access to different levels of water supply
- Indicator 2 (access to sanitation)—the percentage of rural people with access to different levels of sanitation services
- Indicator 3 (health impacts)— disability-adjusted life years (DALYs) for the impacts of WASH services
- Indicator 4 (affordability)—the percentage of household consumption needed to afford safely managed WASH services

Both indicators 1 and 2 take into account the service ladder developed by the United Nations Children's Fund–World Health Organization methodology, from basic access to safely managed.[19]

This section summarizes the KD1 assessment results. For more background and details, refer to the full KD1 report.[20] Appendix A.3 summarizes the KD1 scoring approach and the KD1 scores of the four indicators for all 49 ADB members.

---

[19] Joint Monitoring Programme. 2019. *Progress on Household Drinking Water, Sanitation and Hygiene 2000-2017: Special Focus on Inequalities.* New York: United Nations Children's Fund (UNICEF) and World Health Organization (WHO).

[20] ADB. Forthcoming. KD1 Rural Household Water Security – Final Report. International WaterCentre, Griffith University.

Key Dimension 1
Results across the Regions

Figure 7 shows the population-weighted average KD1 scores of ADB regions. As expected, the Advanced Economies group has the highest water security for rural households, with high scores for all indicators. It is highest of all regions for water supply and sanitation, but not for health impacts and affordability, for which East Asia scores higher. This shows that health and affordability are difficult targets to achieve both in advanced and developing economies. South Asia scores reasonably well on water supply but rather poorly on the other three indicators, resulting in an overall low score. Central and West Asia performs reasonably well on water supply but rather poorly on health impacts and affordability. East Asia receives poor scores for water supply and sanitation but high scores for health impacts and affordability, resulting in the

overall good score. Finally, the Pacific performs poorly for health impacts, likely because the tropical climate can further exasperate some water-related diseases. Results have shown that more affordable water services can improve water and sanitation access, enhancing health outcomes.

Figure 8 shows a positive correlation between the four indicators of KD1. At the horizontal axis the sum of the raw scores of the access to water supply and sanitation is given (indicators 1 and 2). The left vertical axis is the health impacts (indicator 3) and the right vertical axis is the affordability (indicator 4). Correlation is not necessarily causation. Therefore, the conclusion that more affordable water services improve water and sanitation access, enhancing health outcomes, is supported but not proven. The R^2 values for affordability is 0.54 and for health impacts 0.37, both relatively low, suggesting that these are not the only factors.

Figure 7: Population-Weighted Average Key Dimension 1 Results of ADB Regions, 2020

	Water Supply	Sanitation	Health Impacts	Affordability	KD1 Score
Central and West Asia	2.8	1.9	1.7	1.4	7.8
East Asia	2.0	2.0	5.0	5.0	14.0
Pacific	1.3	1.3	1.2	1.5	5.3
South Asia	3.0	1.0	1.2	1.2	6.4
Southeast Asia	2.7	2.2	2.7	3.2	10.8
Advanced Economies	4.9	4.9	4.4	4.8	19.1
Asia and the Pacific*	2.6	1.7	2.8	2.9	10.0

* Without Advanced Economies.

Notes: Maximum score for KD1 is 20; numbers may not sum precisely because of rounding.

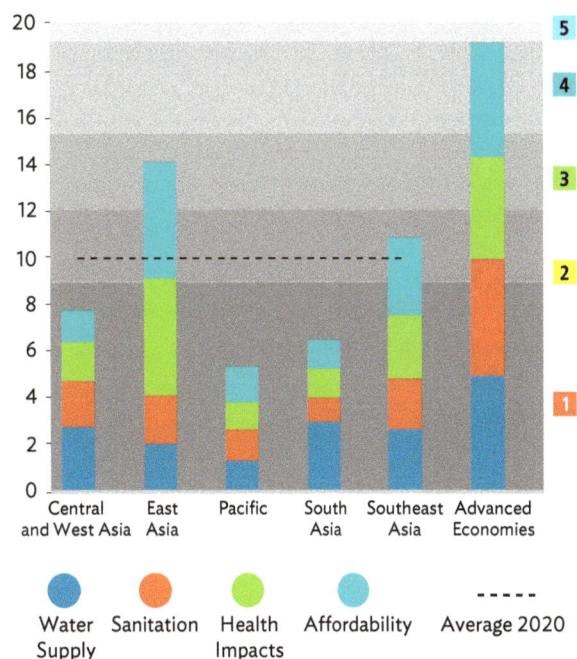

Source: Asian Development Bank.

Figure 8: Correlation Analysis between Indicators, 2020

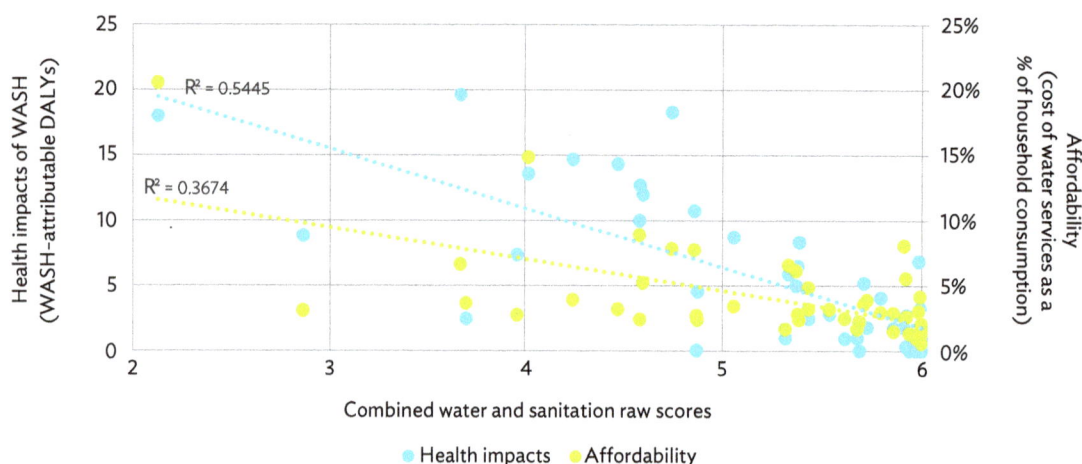

$R^2 = 0.5445$

$R^2 = 0.3674$

Health impacts of WASH (WASH–attributable DALYs)

Affordability (cost of water services as a % of household consumption)

Combined water and sanitation raw scores

● Health impacts   ● Affordability

DALY = disability-adjusted life year; WASH = water, sanitation, and hygiene.
Source: Asian Development Bank.

Overall, the analysis shows good progress on KD1 during 2013–2020.

Good progress: Maldives and Bhutan

Maldives. Most improvement in their water and sanitation access, as they now report more than 99% access to basic water supply and sanitation in rural areas

Bhutan. High scores for water supply and sanitation access, resulting in improved scores for WASH-related health impacts

Risks to Future Rural Household Water Security

Based on the definition of risk, "Risk = Exposure or Hazard x Vulnerability x Coping Capacity," the following main elements of KD1 risk are identified:

- Exposure or hazard—competing users (due to increasing demand), availability of water (due to climate change)

- Vulnerability—human rights, population in poverty, inclusiveness
- Coping capacity—governance, financial and human resources

Climate change impacts on rural and urban water security cover a broad array of issues from a decline and seasonal change in water availability and glacial melt to droughts. The regions will likely face water shortages due to projected changes in climate and growing water demand from rapid population and economic growth.

A closer analysis of these risk elements for the six regions reveals that the main risks for KD1 are

- increased water stress in Central and West Asia and East Asia;
- severe water scarcity in Central and West Asia, East Asia, India, and Bangladesh;
- lack of financial and human resources across all regions; and
- poverty pockets in Central and West Asia and the Pacific.

Rural Household Water Security and Gross National Income

The relation between the KD1 scores of the ADB members (at a scale of 1 to 20) and their GNI per capita is provided in Figure 9. The GNI is plotted at a logarithmic scale. As expected, the graph shows a strong correlation but with some variations. Several ADB members score better than expected considering their GNI, particularly ADB members that belonged to the former Soviet Union. Brunei Darussalam scores rather low in the Advanced Economies group. At the lower GNI scale, ADB members with relatively low scores are Papua New Guinea, Timor-Leste, India (Box 3), the Lao People's Democratic Republic, and Mongolia.

Box 4: Rural Household Water Security in India

India's relatively low score on rural house water security (KD1) does not reflect the major progress the country has made in improving rural drinking water supply and sanitation. For example, the historic sanitation program called Swachh Bharat Mission, initiated in 2014, has resulted in the largest raw score improvement of any country in the Asian Water Development Outlook (AWDO) during 2014–2017. Due to the low level of rural sanitation, the remarkable progress was insufficient to meet the threshold to increase the sub-indicator sanitation to the next level. While this progress is not reflected in the 2020 score yet, it will be shown in the next AWDO.

KD = key dimension.
Source: Asian Development Bank.

Figure 9: Rural Household Water Security and Gross National Income per Capita, 2020

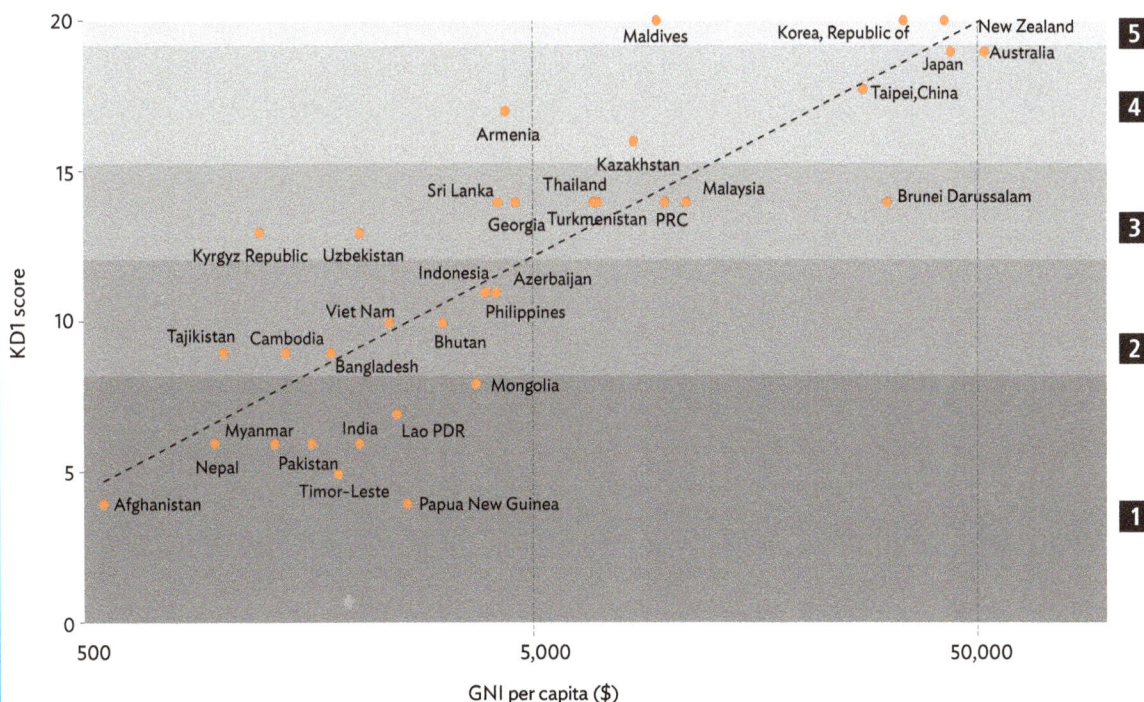

GNI = gross national income, KD1 = key dimension 1 (rural household water security), Lao PDR = Lao People's Democratic Republic, PRC = People's Republic of China.

Note: Small Pacific island states and urban ADB members (Singapore and Hong Kong, China) are not included. The score of India in this graph might be underestimated as the available data for AWDO 2020 did not include yet the results of the extensive projects on rural water supply and sanitation as mentioned in Box 3.

Source: Asian Development Bank.

Gender and Vulnerable People

Women and girls are often tasked with fetching water in rural and peri-urban areas. Without proper sanitation facilities, women and girls may not use the toilet during the day. The burden to fetch water and the lack of access to hygienic, private, and safe sanitation facilities may lead to increased absenteeism and school dropouts, significantly impacting women's contributions to society.

Women's participation in decision-making is significantly lower than men's. Globally, only 24% of lawmakers are women, which is even lower in Asia and the Pacific. Women are not a homogenous group, and their disadvantage varies greatly. Thus, a one-size-fits-all plan without women's participation will not solve the issue. Institutions need to consider inclusive dialogues with all groups when developing WASH policies and plans to ensure that all groups are represented.

Governance and Finance

Poor governance and insufficient funding present risks to the rural household water security (KD1) of ADB members, the majority of which do not have adequately funded and resourced rural water and sanitation plans. Of the 31 ADB members that responded to the 2019 Global Analysis and Assessment of Sanitation and Drinking-Water (GLAAS) survey, only two had allocated the required human and financial resources to complete their water and sanitation policies.[21] This low number is similar globally, with under 10% of ADB members having sufficiently resourced plans. This

shortfall is especially concerning, as it is often poor implementation rather than bad policy that causes "failure to provide adequate WASH services to the poor and other marginalized groups."[22] Also, the Organisation for Economic Co-operation and Development (OECD) survey (explained in the section on Improving Water Security and Key Dimension Performance by Good Governance and Figure 10) showed that WASH policies in the region do not clearly indicate the resources needed to achieve their goals, which hampers their implementation.

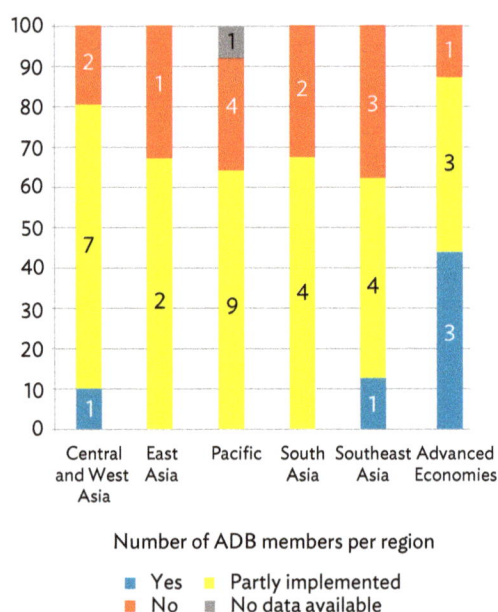

Figure 10: Level of Implementation of Dedicated Water, Sanitation, and Hygiene Policies per Region, 2020 (%)

Number of ADB members per region

■ Yes   ■ Partly implemented
■ No   ■ No data available

Source: Organisation for Economic Co-operation and Development.

---

[21] World Health Organization. 2019. *National Systems to Support Drinking-Water, Sanitation and Hygiene: Global Status Report 2019.* UN-Water Global Analysis and Assessment of Sanitation and Drinking-Water (GLAAS) 2019 Report. Geneva.

[22] World Bank. 2017. *Reducing Inequalities in Water Supply, Sanitation, and Hygiene in the Era of the Sustainable Development Goals: Synthesis Report of the WASH Poverty Diagnostic Initiative.* WASH Synthesis Report. Washington, DC.

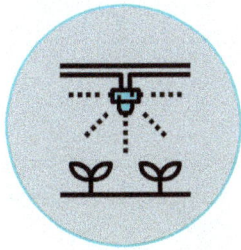

ECONOMIC WATER SECURITY

KD2 assesses the assurance of adequate water to sustainably satisfy a country's economic growth and avoid economic losses due to water-induced disasters.

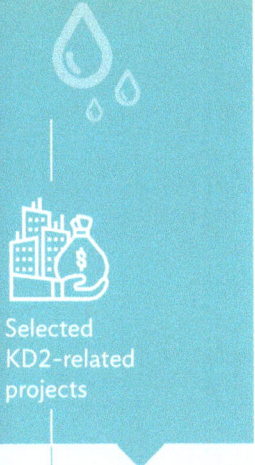

Selected KD2-related projects

Indonesia
Integrated Participatory Development and Management of Irrigation Program

The program will advance the government's food security and rural poverty reduction agenda through increased and improved water delivery. Innovations include a web-based and geospatially interfaced irrigation asset management information system, aerial surveys for improved asset registry, and satellite data for water accounting.

Pakistan
Jalalpur Irrigation Project

The project will build a new surface irrigation system covering about 68,263 hectares of lower productivity, predominantly rainfed land in Punjab province. Satellite remote sensing technology will be incorporated in the monitoring and evaluation of the system for efficient assessment of irrigation efficiency, crop growing, and water productivity.

Indicators included in KD2

Broad economy — scarcity, data availability, reliability, resilience

Agriculture — productivity, self-sufficiency, nutrient security

Energy — productivity, self-sufficiency, energy security

Industry — productivity, self-sufficiency, industry security

KD2 direct links to the SDGs

SDG 2 — TARGET 2.3: By 2030, double the agricultural priority.

SDG 7 — TARGET 7.2: By 2030, increase substantially the share of renewable energy.

SDG 6 — TARGET 6.3: By 2030, substantially increase water-use efficiency across sectors; substantially reduce the number of people suffering from water scarcity.

SDG 8 — Promote, sustained, inclusive, and sustainable growth, and full and productive employment.

What do the numbers tell us about KD2?

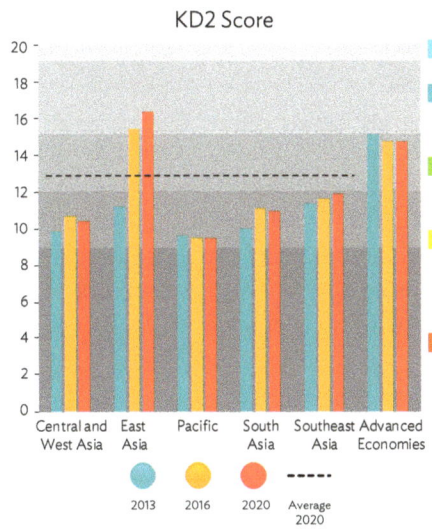

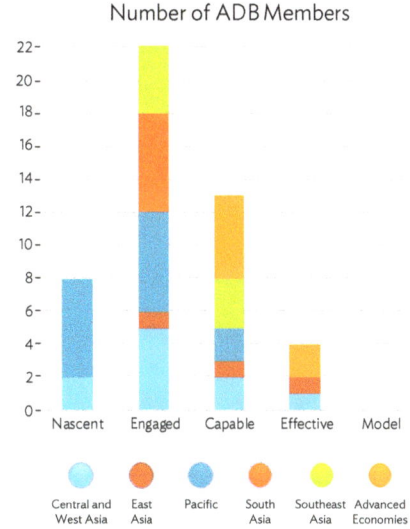

KD2 Score

Categories: Central and West Asia, East Asia, Pacific, South Asia, Southeast Asia, Advanced Economies

Legend: 2013, 2016, 2020, Average 2020

Number of ADB Members

Categories: Nascent, Engaged, Capable, Effective, Model

Legend: Central and West Asia, East Asia, Pacific, South Asia, Southeast Asia, Advanced Economies

East Asia experiences a major increase in their economic water security; the other regions also show progress, although at a slower pace, with the exception of the Pacific

8 ADB members are still in the nascent development stage, 6 in the Pacific region and 2 in Central and West Asia

22 ADB members are still in the engaged development stage with its 2.1 billion people facing restrictions in water supply to support their economic activities

Top performers and challenged ADB members on KD2

Between 2013 and 2020 (scores up at scale of 20)

+5 points
People's Republic of China

+4 points
Uzbekistan

+2 points
Bangladesh • Cook Islands • Turkmenistan

Lower score in 2020

-2 points
Cambodia • Timor-Leste

-1 point
Afghanistan • Australia • Azerbaijan • Japan

Key Dimension 2: Economic Water Security

Introduction

Water is a factor of production needed by every sector of the economy to generate economic growth and development. It must be delivered in the right quantity and quality in the right place at the right time to meet demands for its use in economic production in different sectors. If water cannot be delivered reliably, economic production may be constrained. Ensuring human and environmental health and sustaining reliable water delivery and, therefore, economic production over time, requires water to be set aside for these purposes. Water-related disasters, such as floods, droughts, and landslides, are additional risks to economic production. These risks represented in KD5 (water-related disaster security), the economic impacts imposed, and the costs of mitigating the risks also affect economic water security. KD2 (economic water security) combines these components for AWDO 2020.

KD2: Economic Water Security

Economic water security is a measure of the assurance of adequate water to sustainably satisfy a country's economic growth and avoid economic losses due to water-induced disasters.

KD2 is based on a logical construct of what might constitute a consistent and comprehensive assessment of economic water security. The logic guiding the selection of economic water security indicators for AWDO 2020 is represented in Figure 11 as a decision tree with questions that help determine whether a region may be economically water secure. The first question asks if the region is physically water scarce. If yes, then the region is also economically water scarce, since the volume of water naturally available in the region is not enough to meet the demands of economic production. Alleviating physical water scarcity is possible through importing water from another region. Trade is placed in a box by itself on the right of Figure 11 to indicate that it is an external supply factor that can impact water security across all categories. Desalination can also alleviate physical and economic water scarcity and provide an opportunity to export water to the region.

If the region is not physically water scarce, the next question for assessing economic water security is determining whether water supply is reliable and risk management robust so that water is being provided reliably and affordably in adequate quantity and quality to meet all demands, including environmental demands. If yes, economic water security is maximized. If no, then the region is not entirely economically secure and can investigate various components of water security to determine the degree to which water is secured. Within this framework, the extent to which measures have been taken to secure water for economic use must be considered.

Interventions to improve economic water security in the regions include enhancing water storage capacity, managing water-related risks, optimizing water allocation, and improving water-use efficiency and productivity. Each method must be economically feasible for water to be economically secure. If the solution cost exceeds the value generated by using the water, water cannot be considered economically secure because it cannot be sustainably provided. KD2 does not include the losses caused by drought and flood disasters, which are addressed in KD5.

KD2 indicators are

- broad economy (availability, reliance, etc.);
- agriculture (productivity, self-sufficiency, nutrient security);
- energy (productivity, self-sufficiency, energy security); and
- industry (productivity, self-sufficiency, industry security).

Figure 11: Economic Water Security Construct

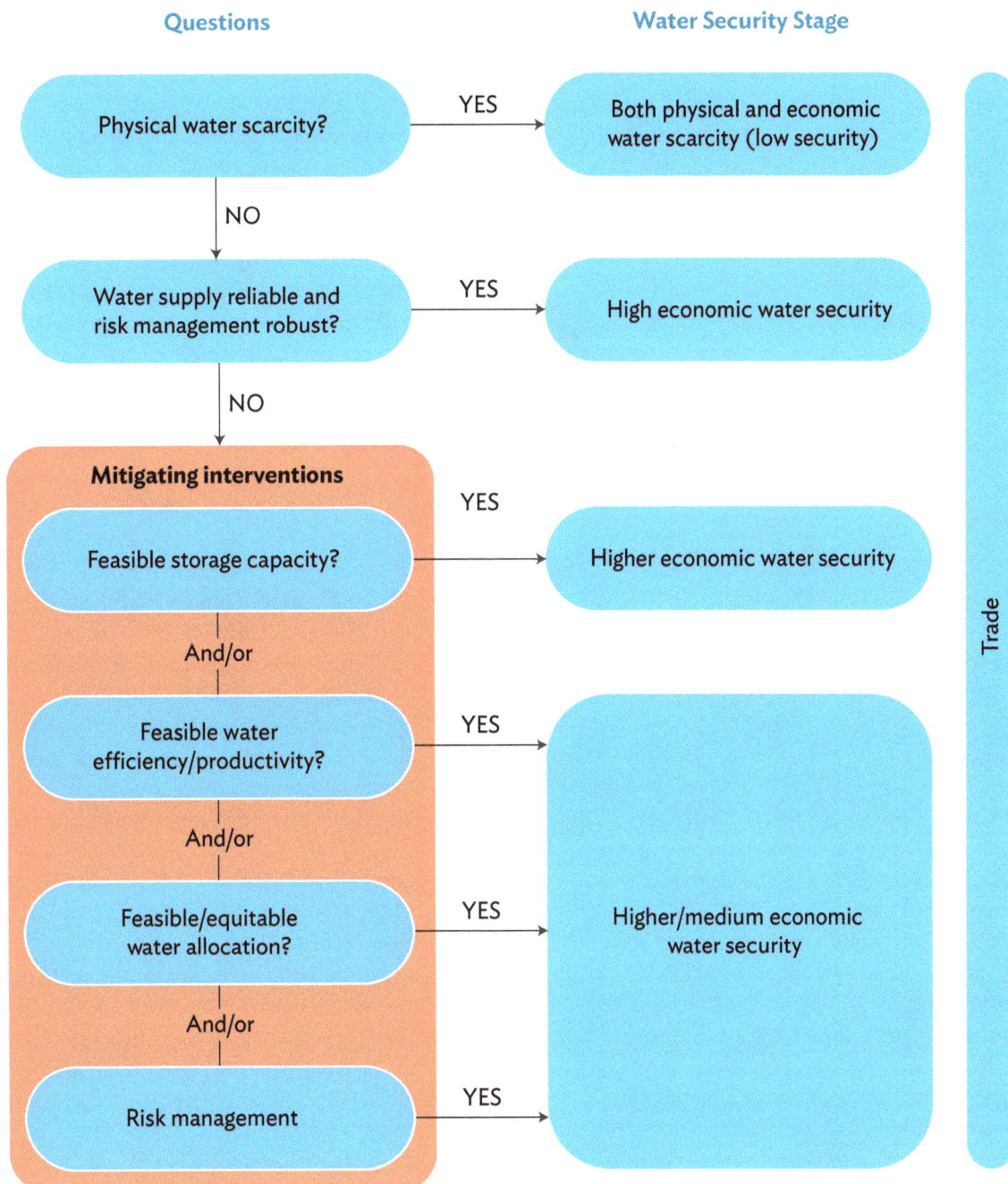

Questions

Water Security Stage

Physical water scarcity? — YES → Both physical and economic water scarcity (low security)

NO ↓

Water supply reliable and risk management robust? — YES → High economic water security

NO ↓

Mitigating interventions

Feasible storage capacity? — YES → Higher economic water security

And/or

Feasible water efficiency/productivity? — YES →

And/or

Feasible/equitable water allocation? — YES → Higher/medium economic water security

And/or

Risk management — YES →

Trade

Source: Asian Development Bank.

Food security is not addressed in AWDO. In many countries, food security strongly relies on rainfed agriculture. However, achieving food security requires much more than water. The sub-indicator nutrient security in agriculture provides a score on the additional calories per capita needed to reach the average consumption in Asia and the Pacific. The indicator agriculture includes the productivity and self-sufficiency of the irrigation part of food production and, as such, demonstrates water efficiency in food production.

This section summarizes the results of the KD2 assessment. For more details, please see the full KD2 report.[23] Appendix 4 details the scoring approach of KD2 and the KD2 scores of the four indicators for all 49 ADB members.

Key Dimension 2 Results across the Regions

Figure 12 shows the population-weighted average KD2 scores of ADB regions. Some of the results may initially seem surprising. ADB members that rank highly are spread across different regions of Asia (see also Appendix 4). The most advanced ADB members are not always shown to be the most economically water secure, and some may be relatively dry. ADB members are shown to be comparatively secure. These results fit the definition of economic water security, which does not represent physical water scarcity but is a measure of the assurance of adequate water to sustainably satisfy a country's economic growth and accommodate economic losses due to water-induced disasters.

Some ADB members with lower quantities of renewable water resources might also have lower demands, so less water is needed. ADB members may have developed the infrastructure necessary to reliably deliver water to meet demands, reduce risks, and use water more productively. Kazakhstan, for example, ranks second in its economic water security

score, despite having less annual precipitation than most other ADB members. However, Kazakhstan demands a smaller share of its total water resources than other ADB members; has comparatively high water storage capacity; can be self-sufficient in agriculture, energy, and industry production; is comparatively nutrient secure; and is above average in electricity and industrial production per capita. These factors indicate that water scarcity is not currently limiting economic development. Therefore, the mitigating factors provide the potential for Kazakhstan to be more economically water secure, even with limited renewable water resources. The high scores of Taipei,China lead to East Asia's high overall score, which is higher than the score of the Advanced Economies.

Risks to Future Economic Water Security

The main future risk for KD2 is climate change, which affects water productivity, reliability, and self-sufficiency in various sectors. While climate change is generally anticipated to result in increased annual and seasonal precipitation over most regions of Asia, it is also expected to increase variability with wet periods getting wetter and dry periods getting dryer. The extent to which this will translate into greater or more reliable water supplies for agriculture and communities will depend on rainfall patterns; the effective management of existing reservoirs, including groundwater, wetlands, and soil moisture; and the construction of new storage to smooth the annual discharge cycle, providing water during the dry periods for water supply and irrigation. The same applies to energy water security and, to a lesser extent, industrial water security. ADB members should prepare themselves to adapt to climate change and invest in climate-resilient measures.

Another risk for KD2 might result from transboundary issues. Growing populations and increased socioeconomic activities will lead to

---

[23]   ADB. Forthcoming. KD2 Economic Water Security – Final Report. International Water Management Institute.

Figure 12: Population-Weighted Average Key Dimension 2 Results of ADB Regions, 2020

	Broad Econ	Agriculture	Energy	Industry	Score
Central and West Asia	2.1	3.0	2.8	2.6	10.4
East Asia	3.0	4.3	4.0	5.0	16.3
Pacific	2.8	1.7	2.5	2.6	9.5
South Asia	1.9	3.4	2.7	3.0	10.9
Southeast Asia	2.4	3.6	2.8	3.0	11.9
Advanced Economies	2.5	3.1	4.4	4.9	14.8
Asia and the Pacific*	2.4	3.7	3.2	3.7	13.0

* Without Advanced Economies.

Notes: Maximum score for KD2 is 20; numbers may not sum precisely because of rounding.

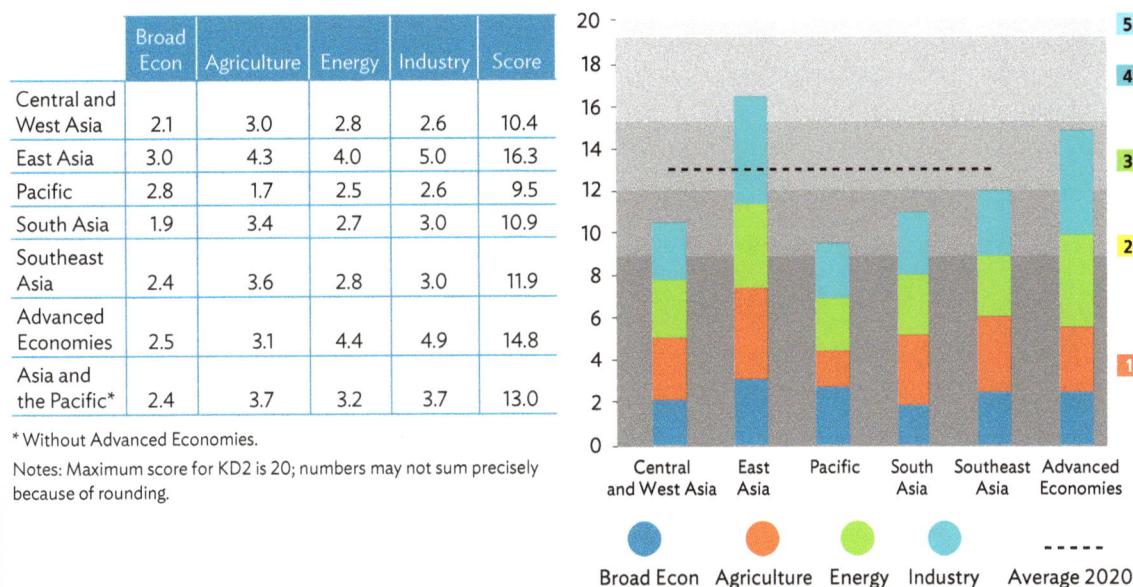

Source: Asian Development Bank.

increased demands and withdrawals. In case this takes place upstream, less water will be available downstream. It might also result in increased salinity intrusion from the sea in downstream regions. Furthermore, upstream pollution might constrain water use downstream. Basin-wide agreements between the riparian regions are needed to control these risks. Therefore, basin-wide water resources management and regional cooperation will be instrumental in mitigating transboundary risks.

Economic Water Security and Gross National Income

The relation between the KD2 scores of ADB members (at a scale of 1–20) and their GNI per capita is provided in Figure 13. The GNI is plotted at a logarithmic scale. The correlation is similar to KD1 with a clear relation (as expected) but with a large variation. On the positive side are the PRC and Kazakhstan. Bangladesh is also performing well despite its low GNI.

Governance and Economic Water Security

The main condition for economic water security is effective governance, which entails developing and managing the water resource system to support the economic sectors. Investments in infrastructure are needed to store and distribute water. Robust management is also essential to ensure equitable water allocation. The OECD framework on water governance (see section on Improving Water Security and Key Dimension Performance by Good Governance) includes the following principles that are particularly important for KD2: Principle 5 (data and information), Principle 7 (regulatory frameworks), Principle 10 (stakeholder engagement), Principle 11 (trade-offs), and Principle 12 (monitoring and evaluation). Two of section III's policy recommendations directly address these governance requirements on monitoring and integrated water resources management (IWRM).

Figure 13: Economic Water Security and Gross National Income per Capita, 2020

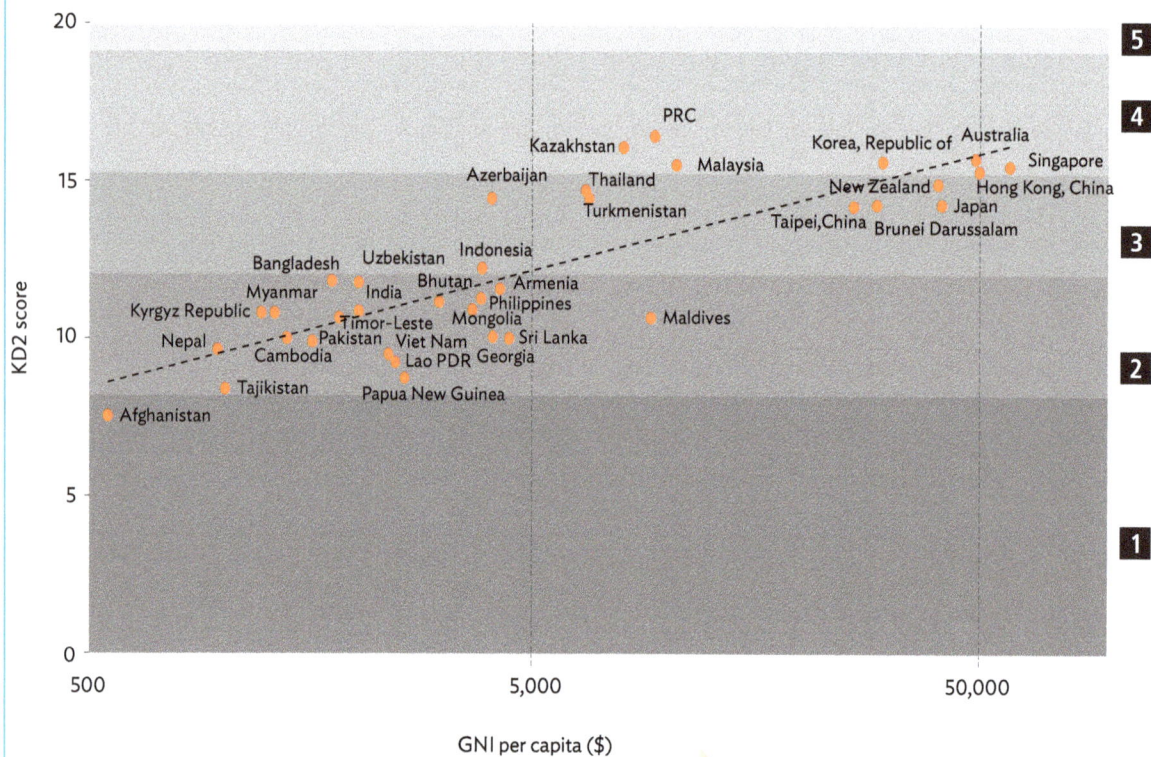

GNI = gross national income, KD2 = key dimension 2 (economic water security), Lao PDR = Lao People's Democratic Republic, PRC = People's Republic of China.

Note: Small Pacific island states are not included.

Source: Asian Development Bank.

KEY DIMENSION 3
URBAN WATER SECURITY

KD3 assesses the extent ADB members are providing safely managed and affordable water and sanitation services for their urban communities to sustainably achieve desired outcomes.

Indicators included in KD3

Access to water supply | Access to sanitation | Affordability | Drainage (flooding) | Environment (water quality)

KD3 direct links to the SDGs

SDG 6 — **TARGET 6.1:** By 2030, achieve universal and equitable access to safe and affordable drinking water for all.

TARGET 6.2: By 2030, achieve access to adequate and equitable sanitation and hygiene for all, and end open defecation, paying special attention to the needs of women and girls and those in vulnerable situations.

SDG 11 — Make cities and human settlements inclusive, resilient, and sustainable.

What do the numbers tell us about KD3?

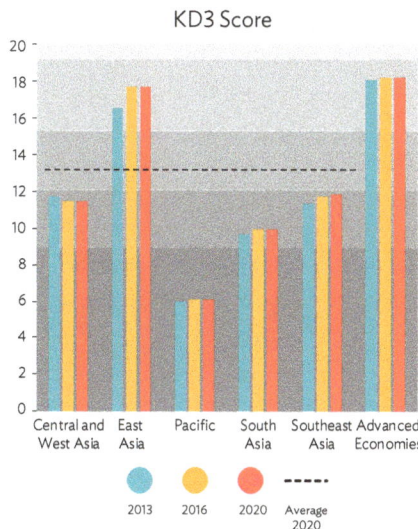

East Asia and Southeast Asia experience some increase in their urban water security; the other regions do not make much progress ▪

7 ADB members are still in the nascent development stage, 6 in the Pacific (Kiribati, the Federated States of Micronesia, Nauru, Papua New Guinea, Solomon Islands, Vanuatu) and 1 in Southeast Asia (Timor-Leste), affecting 1.9 million people ▪

18 ADB members are still in the engaged development stage with 790 million urban people in these 18 ADB members facing insufficient urban water security ▪

KD3 Score

Categories: Central and West Asia, East Asia, Pacific, South Asia, Southeast Asia, Advanced Economies

Legend: 2013 · 2016 · 2020 · Average 2020

Number of ADB Members

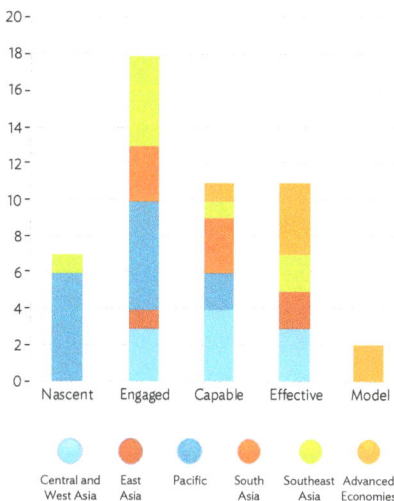

Categories: Nascent, Engaged, Capable, Effective, Model

Legend: Central and West Asia · East Asia · Pacific · South Asia · Southeast Asia · Advanced Economies

Top performers and challenged ADB members on KD3

Between 2013 and 2020 (scores up at scale of 20)

+4 points — Palau

+2 points — Lao PDR • Marshall Islands

Lower score in 2020

−1 point — Federated States of Micronesia • Nepal • Pakistan • Solomon Islands • Vanuatu

---

Selected KD3-related projects

Bangladesh
Dhaka Water Supply Network Improvement

The proposed project will improve the water supply system in Dhaka by making it more reliable, sustainable, and climate-resilient. Innovative features include trenchless technology for pipe rehabilitation, district metered areas, and use of supervisory control and data acquisition (SCADA) systems.

Sri Lanka
Greater Colombo Water and Wastewater Management Improvement Investment Program

The project will rehabilitate and expand water supply and wastewater management infrastructure in Greater Colombo. Innovative features include trenchless technology, district metered areas and a wastewater treatment plant constructed through a Design-Build-Operate contract.

Key Dimension 3: Urban Water Security

Introduction

Urban water security means sustainably meeting the community's water needs, which are (i) agreed in technical, economic, environmental, and social dimensions; and (ii) met now and in the future. Operationalizing urban water security by integrating with the SDGs for water and sanitation leads to the functional definition of KD3 in AWDO 2020.[24]

KD3: Urban Water Security

Urban water security assesses the extent to which countries are providing safely managed and affordable water and sanitation services for their urban communities to sustainably achieve desired outcomes.

KD3 is a composite of five indicators:

- water supply (service ladder standards),
- sanitation (service ladder standards),
- affordability,
- drainage (urban flooding), and
- environmental water security.

Globally, more people live in urban areas than in rural areas. With cities rapidly growing and often also the centers of economic productivity, the importance of urban water security is increasing. The rate of urbanization in Asia was approximately 2.2% per annum during 2015–2020. Although the rate is projected to decline to 0.8% by 2045, urban growth is expected into the future. Asia has a relatively low level of urbanization at about 50%

of the population compared with a global average of about 55%. However, in real terms, the region is home to the largest urban population in the world. Providing urban water security is essential in achieving sustainable, livable, resilient, and productive cities. But continuing urban growth and climate change impacts create significant challenges to providing water, sanitation, and stormwater infrastructure.

This section summarizes the KD3 assessment results. For more details, please see the full KD3 report.[25] Appendix 5 details the scoring approach of KD3 and the KD3 scores of the five indicators for all 49 ADB members.

Key Dimension 3 Results across the Regions

While many ADB members with Advanced Economies—such as Hong Kong, China; New Zealand; and Singapore—can achieve and maintain high urban water security, most Pacific developing member countries (DMCs) continue to face challenges. Low urban water security is particularly evident in the Federated States of Micronesia, Papua New Guinea, Kiribati, and Nauru (Pacific) and Timor-Leste (Southeast Asia). These countries are the only ADB members to have received an urban water security index of 1. A very low proportion of their urban population receives affordable, safely managed water and sanitation services, and is subject to high economic impacts of floods and storms.

Figure 14 provides the population-weighted average KD3 scores of ADB regions, showing the development during 2013, 2016, and 2020. The Pacific, South Asia, Southeast Asia, East Asia, and Advanced Economies all see a gradual improvement in their urban water security scores over time. East Asia particularly stands out, mostly attributed

---

24   See also Allan, J. V., S. J. Kenway, and B. W. Head. 2018. Urban Water Security - What Does It Mean? *Urban Water Journal*. 15 (9). pp. 899–910.

25   ADB. Forthcoming. KD3 Urban Water Security – Final Report. International WaterCentre, Griffith University, University of Queensland.

Figure 14: Population-Weighted Average Key Dimension 3 Results of ADB Regions, 2020

	Water Supply	Sanitation	Affordability	Drainage	Environment	KD1 Score
Central and West Asia	3.6	1.9	4.8	0.8	0.4	11.5
East Asia	5.9	5.9	4.7	0.9	0.6	17.9
Pacific	1.4	1.3	1.6	1.0	0.9	6.3
South Asia	2.7	1.2	4.7	0.9	0.6	10.1
Southeast Asia	3.0	2.0	5.3	0.8	0.9	11.9
Advanced Economies	5.9	5.9	4.8	0.9	1.0	18.4
Asia and the Pacific*	3.9	3.1	4.8	0.9	0.6	13.3

* Without Advanced Economies.

Notes: Maximum score for KD3 is 20; numbers may not sum precisely because of rounding.

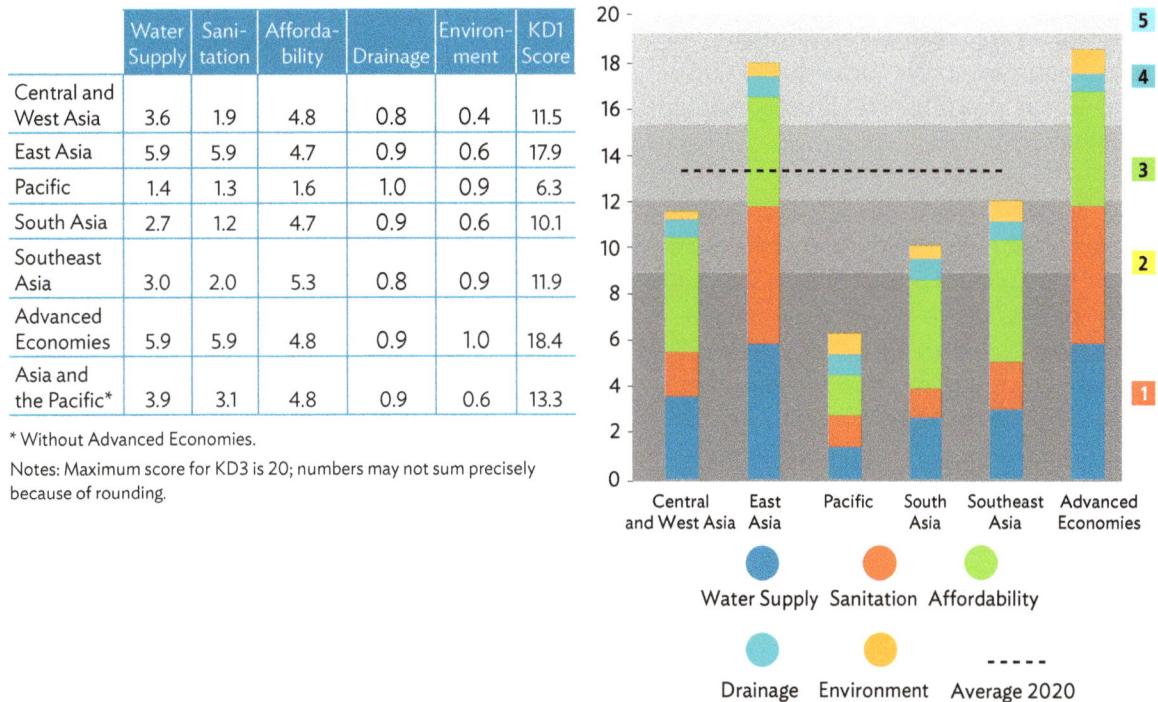

Water Supply   Sanitation   Affordability   Drainage   Environment   Average 2020

Source: Asian Development Bank.

to increases in the coverage of safely managed sanitation services in the PRC. Central and West Asia is the only region with a notable decrease in the urban water security score due to a single-point reduction in Pakistan's water supply index.

Risks to Future Urban Water Security

In its previous editions, AWDO focused primarily on infrastructure-related urban water security issues such as piped water services and sewage collection networks. In 2020, AWDO built on this solid foundation and incorporated natural, technical, social, financial, and institutional risks into future urban water security. Four risk sub-indicators are quantitatively assessed: urban growth rate (%), nonrevenue water (%), water consumption (liter per person per day [l/p/d]), and energy cost (%).

The current indicator value and its history are considered in order to reflect the likelihood and consequence of high-risk events in the future.

The urban growth rate is a particularly important influence on urban water risk because growing populations need more infrastructure and servicing. Seven ADB members have urban population growth rates of over 3%. Nonrevenue water indicates water and asset management, the need for new supplies, and the chance of loss should new supplies be added. This indicator ranged from about 4% for Singapore to over 40% for Azerbaijan and Pakistan. While water consumption is expected to vary according to geographic and climatic conditions, using high volumes may be wasteful, while using too little may reflect poor access. The analysis shows extreme values from 35 l/p/d in Afghanistan to 308 l/p/d in Australia. Energy can be a significant cost component of water service provision. Access to low-cost, reliable energy supplies can support improved water treatment outcomes and a shift to more climate-resilient water supplies (such as recycled water or seawater desalination). This indicator ranged from 5.6% in the PRC to around 40% in India,[26] Pakistan, Bangladesh, and Afghanistan.

Ten ADB members have sufficient data for all four indicators (Figure 15). Pakistan and Afghanistan have the highest levels of risk to future security. The Philippines, Viet Nam, Uzbekistan, Indonesia, and the PRC show low levels of risk to future urban water security. The limited number of ADB members in this assessment suggests that additional data should be required to depict future urban water security risks more completely. As more data become available, more sub-indicators could be included.

Climate change and competition for water are additional risk factors to be taken into account in KD3. Limited availability of good quality surface water and groundwater for urban water supply may require expensive infrastructural solutions. The rapid decrease in groundwater levels is a major concern in some urban areas that rely on groundwater resources.

Urban Water Security and Gross National Income

Figure 16 shows the relation between the KD3 scores of ADB members (at a scale of 1–20) and their GNI per capita, which is much weaker than for KD1 and KD2. The GNI is plotted at a logarithmic scale. As expected, the Advanced Economies perform well, except Brunei Darussalam. The PRC, the Philippines, the Lao PDR, Uzbekistan, and the Kyrgyz Republic show good scores. In contrast, Thailand and Malaysia receive low scores.

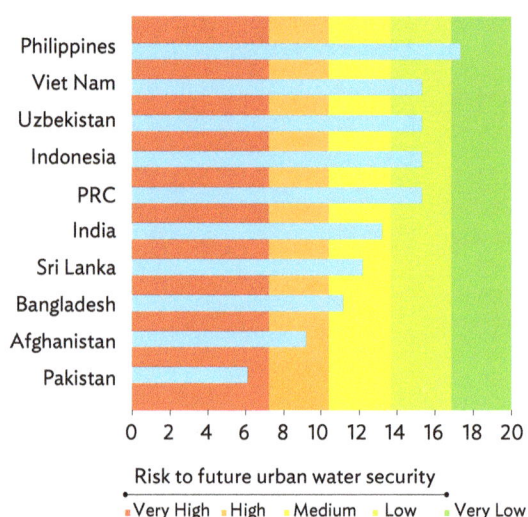

Figure 15: Future Urban Water Security Risk Scores

PRC = People's Republic of China.
Source: Asian Development Bank.

Governance

Effective governance arrangements are critical to successfully implementing water security solutions and delivery over the short, mid-, and long term. Wider consideration of the role and scope of governance appears to be a high priority and relatively under-assessed area, particularly considering policies, laws, rights, and the political and institutional enabling environment.

[26]   The International Benchmarking Network. https://www.ib-net.org/about-us/about-ibnet/ (accessed 27 October 2020).

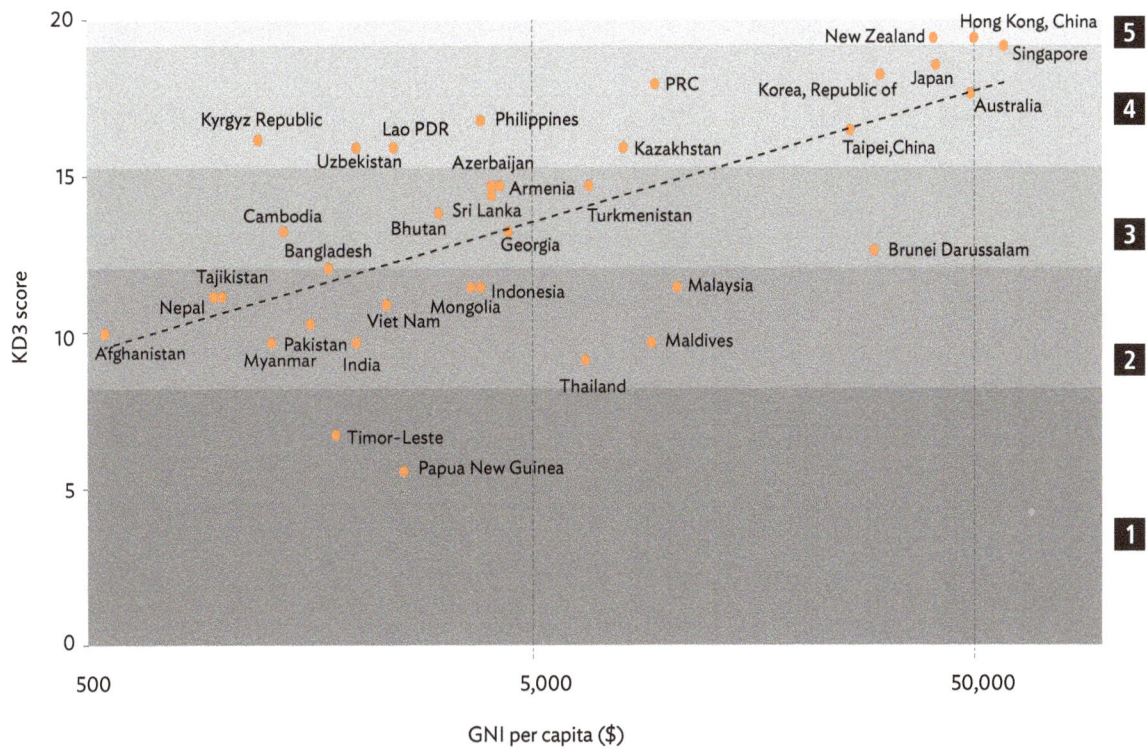

Figure 16: Urban Water Security and Gross National Income per Capita, 2020

GNI = gross national income, KD3 = key dimension 3 (urban water security), Lao PDR = Lao People's Democratic Republic, PRC = People's Republic of China.

Notes: The scores of India and Thailand in this graph might be underestimated due to missing data on safely managed water supply and sanitation. See Appendix 5 for further information.
    Small Pacific island states are not included.

Source: Asian Development Bank.

Inle Lake, a freshwater lake located on the Shan Plateau, Myanmar

ENVIRONMENTAL WATER SECURITY

KD4 assesses pressures on the health of rivers, wetlands, and groundwater systems and measures progress in restoring aquatic ecosystems to health on a national and regional scale.

Selected KD4-related projects

Nepal
Bagmati River Basin Improvement Project

The project is the first attempt in Nepal to apply the concept and principles of IWRM since its adoption under the 2005 National Water Plan. It will invest in forming a river basin organization with adequate capacity and decision support systems to enhance water security in the basin.

Lao PDR
Sustainable Rural Infrastructure and Watershed Management

This integrated project will improve rural incomes from market-driven diversified farm output, watershed health, and community nutrition. The project aims to promote diversification of crops, improve yields by providing irrigation and access, and protect watershed ecological services.

Indicators included in KD4

Catchment and Aquatic System Condition Index (CASCI)
land cover and hydrology alteration, groundwater, connectivity, and water quality

Environmental Governance Index (EGI)
sustainable nitrogen management, wastewater treatment, and terrestrial protection

KD4 direct links to the SDGs

SDG 3 Ensure healthy lives and promote well-being for all at all ages.

SDG 6 Ensure access to water and sanitation for all.

SDG 14 Conserve and sustainably use the oceans, seas, and marine resources.

SDG 15 Sustainably manage forests, combat desertification, halt and reverse land degradation, and halt biodiversity loss.

What do the numbers tell us about KD4?

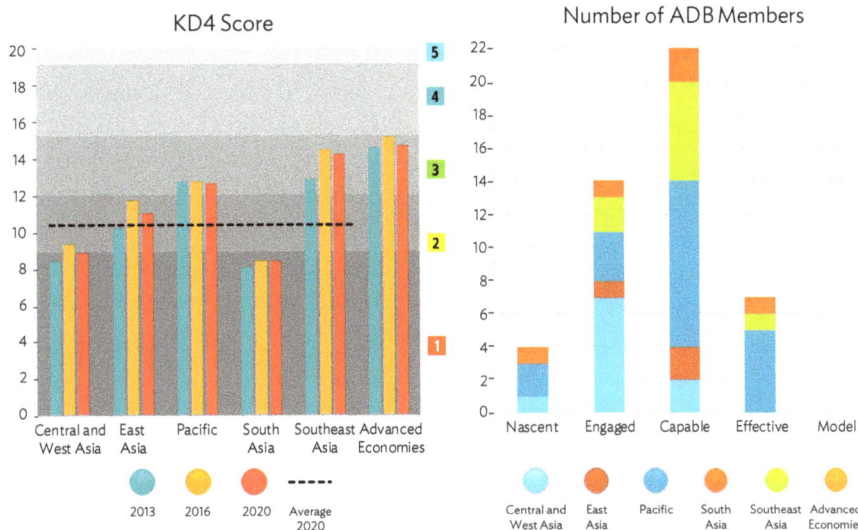

KD4 Score

(Bar chart showing KD4 Score by region for 2013, 2016, 2020 with Average 2020 dashed line at approximately 10.5)

Categories: Central and West Asia, East Asia, Pacific, South Asia, Southeast Asia, Advanced Economies

Legend: 2013, 2016, 2020, Average 2020

Number of ADB Members

(Stacked bar chart by stage: Nascent, Engaged, Capable, Effective, Model)

Legend: Central and West Asia, East Asia, Pacific, South Asia, Southeast Asia, Advanced Economies

Southeast Asia and Pacific ADB members score well; South Asia lags behind

Four ADB members are still the nascent development stage: India, Pakistan, Singapore, and Maldives

No major developments in time; some progress was made between 2013 and 2016; since 2016, some regions have shown a lower environmental water security

Top performers and challenged ADB members on KD4

Between 2013 and 2020 (scores up at scale of 20)

+8 points
Marshall Islands

+5 points
Maldives

+4 points
Philippines •
Taipei,China •
Tuvalu

+3 points
Kyrgyz Republic •
Federated States of
Micronesia • Samoa •
Timor-Leste • Viet Nam

Lower score in 2020

−3 points
Republic of Korea

−1 point
Bhutan • Solomon Islands

Key Dimension 4: Environmental Water Security

Introduction

In Asia and the Pacific, freshwater systems are an integral part of many landscapes and are inextricably linked to human lives. Healthy waterways provide a range of ecosystem services, including good quality water, basic flood protection via natural wetlands, and food security from agricultural products and from healthy fisheries in both freshwater and coastal systems. However, human alteration of the environment negatively impacts the health of aquatic ecosystems. For example, physical changes to the landscape, removal of riparian vegetation, and depletion of groundwater can lead to downturns in water availability, water quality, and biodiversity, along with a weakened resilience to natural disasters. Consequently, human health and well-being and the economy can be negatively affected. Effective assessment of aquatic ecosystem health is thus vital to understanding the environmental water security of Asia and the Pacific.

KD4: Environmental Water Security

Environmental water security assesses the health of rivers, wetlands, and groundwater systems and measures the progress in restoring aquatic ecosystems to health on a national and regional scale.

Two indicators contribute to the KD4 scores, providing complementary assessments of aquatic ecosystem health and governance arrangements to maintain and restore healthy waterways. These two indicators comprise several sub-indicators:

- Catchment and Aquatic System Condition Index (CASCI)
  - » Riparian land cover change
  - » Hydrological alteration
  - » Groundwater depletion
  - » Water quality
  - » Riverine connectivity
- Environmental Governance Index (EGI)
  - » Wastewater treatment
  - » Terrestrial protected areas
  - » Sustainable Nitrogen Management Index

This section summarizes the results of the KD4 assessment. For more details, please see the full KD4 report.[27] Appendix 6 lists the scoring approach of KD4 and the KD4 scores of the two indicators for all 49 ADB members.

Key Dimension 4 Results across the Regions

Figure 17 illustrates the overall KD4 population-weighted average scores of ADB regions, showing the development during 2013–2020. Southeast Asia and the Pacific score above average, close to the results of the Advanced Economies, mainly due to high EGI, although some countries have relatively low levels of land cover and flow alteration to moderately high CASCI. While the overall score for KD4 in the Pacific is good, the score for CASCI is low. All regions, except the Pacific, show some progress in KD4 during 2013–2016. The largest overall increase during 2013–2020 is in Southeast Asia, particularly due to the good performance of the Philippines and Viet Nam (Box 5). A slight decline is recorded during 2016–2020. The Republic of Korea has a low score due to a decline in CASCI (Box 6).

---

27   ADB. Forthcoming. KD4 Environmental Water Security – Final Report. International WaterCentre, Griffith University.

Figure 17: Population-Weighted Average Key Dimension 4 Results of ADB Regions, 2020

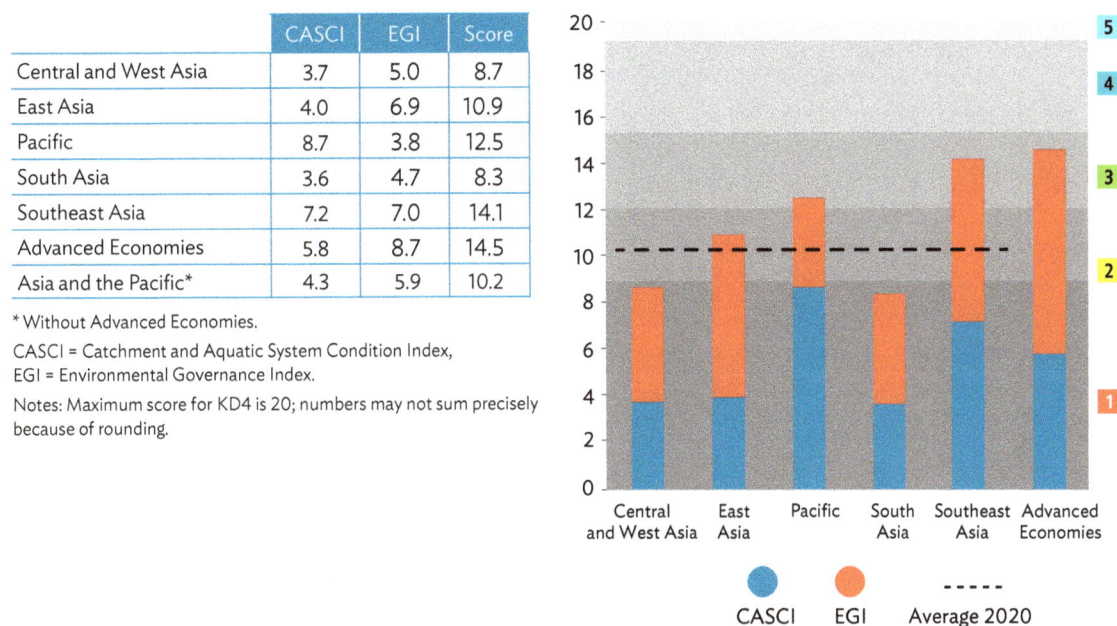

	CASCI	EGI	Score
Central and West Asia	3.7	5.0	8.7
East Asia	4.0	6.9	10.9
Pacific	8.7	3.8	12.5
South Asia	3.6	4.7	8.3
Southeast Asia	7.2	7.0	14.1
Advanced Economies	5.8	8.7	14.5
Asia and the Pacific*	4.3	5.9	10.2

* Without Advanced Economies.
CASCI = Catchment and Aquatic System Condition Index,
EGI = Environmental Governance Index.

Notes: Maximum score for KD4 is 20; numbers may not sum precisely because of rounding.

Source: Asian Development Bank.

Box 5: Good Progress: Viet Nam's Environmental Water Security

During 2013–2020, Viet Nam's key dimension 4 (KD4) (environmental water security) score has increased substantially, primarily due to Environmental Governance Index (EGI) improvements and an increase in terrestrial protected areas, and partly due to wastewater treatment improvements. At the highest levels of the Government of Viet Nam, there have been policy decisions to increase the number of national parks, such as Decision No. 1976/2014/QD-TTg of the Prime Minister dated 30 October 2014 on the approval of the planning for special use forest system to 2020, and a vision to 2030. This decision includes ongoing increases in protected areas, which may increase EGI in future Asian Water Development Outlook (AWDO) reports.

Source: Asian Development Bank.

Box 6: Lagging in Progress: The Republic of Korea's Environmental Water Security

During 2013–2020, the Republic of Korea's key dimension 4 (KD4) (environmental water security) score has decreased due to a decline in the Catchment and Aquatic System Condition Index (CASCI), driven primarily by riparian vegetation alteration. Despite substantial revegetation policies since the 1960s, there has been a small alteration of vegetation in the riparian area around waterways and wetlands since 2000. The reduced KD4 score is also because a relatively large proportion of this riparian vegetation removal has occurred in the last 5 years. This indicates that there appears to be a growing trend of removing vegetation close to the stream, which directly impacts aquatic ecosystem health. Halting this trend while maintaining good performance in other sub-indicators of KD4 will improve the score.

Source: Asian Development Bank.

Risks to Future Environmental Water Security

The use of CASCI and EGI indicators can identify ADB members and regions at risk of future declines in environmental water security, even if they currently have a high CASCI. A country is considered at risk of future declines in environmental water security if it has a high CASCI but a low EGI. Such a country would have comparatively limited pressures on aquatic ecosystem health. However, with limited ecosystem protection and management of wastewater and nutrient pollution, as indicated by the low EGI, it may be at risk of future declines in environmental water security.

ADB members with large differences between CASCI and EGI are in the Pacific: Samoa, the Federated States of Micronesia, Vanuatu, Papua New Guinea, Solomon Islands, and Fiji (Figure 18). Except for Solomon Islands, which has a lower CASCI and EGI, these members have a CASCI score above 4 but an EGI below 2.

Aquatic ecosystems, and consequently environmental water security, will face escalating climate change impacts on many levels. Rising temperature trends will likely disrupt natural flow regimes and riverine connectivity, decrease water volume and quality, and exacerbate direct pressures already faced by instream organisms and riparian vegetation. Shifting rainfall patterns are effectively a form of long-term flow alteration transforming the fundamental hydrology of aquatic systems and leading to decreases in the abundance and diversity of native aquatic organisms, thereby impacting aquatic ecosystem health. Aquatic biodiversity losses will also affect food security, particularly where inland fisheries provide an important source of protein.

Environmental Water Security and Gross National Income

Figure 19 shows the relation between the KD4 scores of ADB members (at a scale of 1–20)

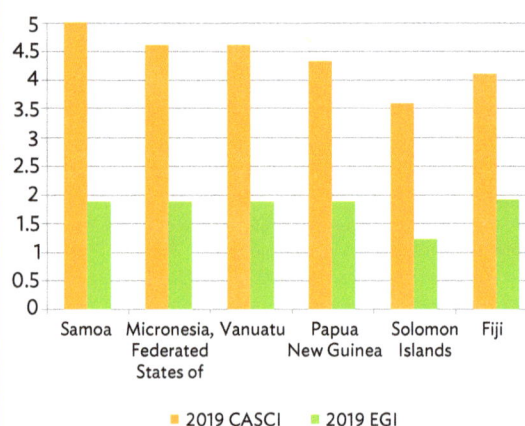

Figure 18: ADB Members with the Largest Differences between Catchment and Aquatic System Condition Index and Environmental Governance Index

CASCI = Catchment and Aquatic System Condition Index, EGI = Environmental Governance Index.
Note: A large difference between CASCI and EGI signifies risk to environmental water security.
Source: Asian Development Bank.

and their GNI per capita. The GNI is plotted at a logarithmic scale. It appears that the correlation is not strong, with considerable variation at both the lower and higher GNI levels. ADB members with less developed economies—Afghanistan, Nepal, and Tajikistan—score reasonably high in KD4, potentially because their environment has not yet been impacted negatively by economic development. In contrast, densely populated developing ADB members, such as Pakistan, India, and Bangladesh, show low scores mainly due to economic growth. As these countries become more developed, they may focus on environmental restoration like wetland preservation, improved groundwater management, and increased riverine connectivity. Ideally, countries will follow a more sustainable path avoiding this stage of environmental degradation during rapid development.

Figure 19: Environmental Water Security and Gross National Income per Capita, 2020

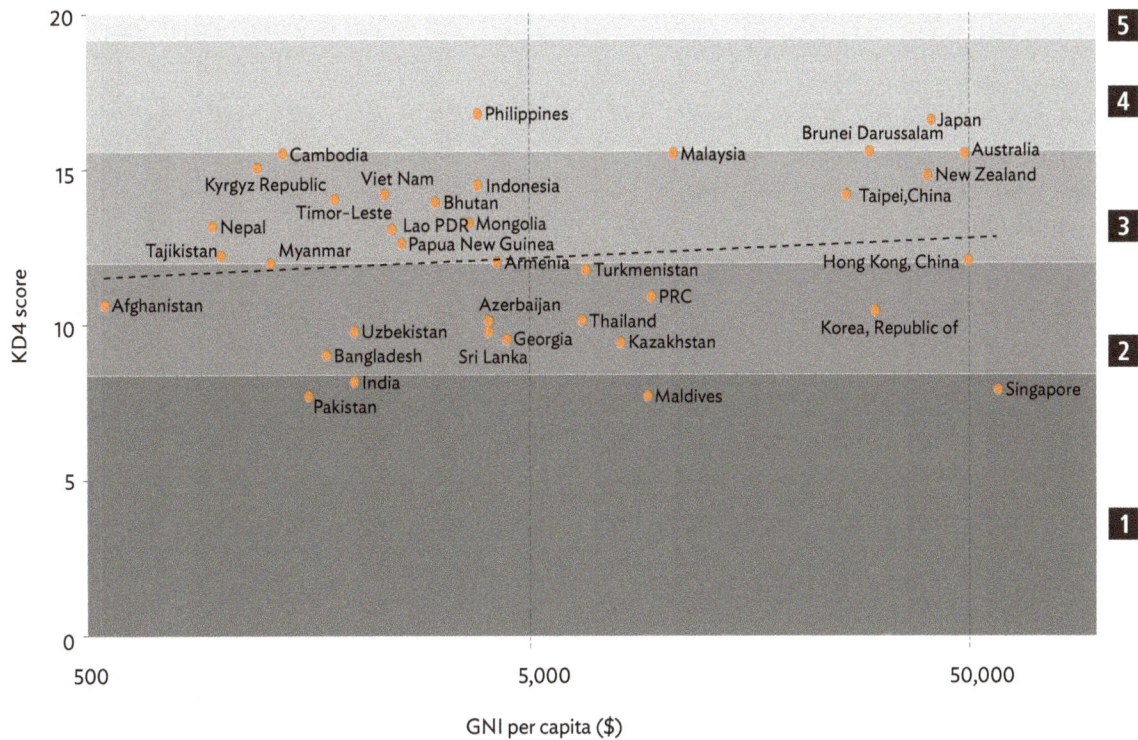

GNI = gross national income, KD4 = key dimension 4 (environmental water security), Lao PDR = Lao People's Democratic Republic, PRC = People's Republic of China.
Note: Small Pacific island states are not included.
Source: Asian Development Bank.

Gender and Vulnerable People

Social equity in relation to aquatic ecosystem health is multidimensional. It includes (i) recognizing all stakeholders and their inclusiveness in decisions surrounding conservation and management, and (ii) distributing costs, benefits, burdens, and rights to the stakeholders. Gender equity is embedded in race, religion, and class. For example, women from the lower class in South Asian countries are involved in water services labor, such as irrigation of crops, watering cattle, and collecting drinking water. However, forms of access to aquatic resources that involve power, ownership of land, technology (pumps), or infrastructure tend to be dominated by men. Improving social and gender equity in obtaining ecosystem services and managing aquatic ecosystems requires identifying and understanding relationships between equity and environmental health globally.

Governance in Key Dimension 4

Governance is one of the two main indicators of KD4. Using EGI in KD4 focuses on how successful governance is in terms of the three components: % of wastewater treated, % of area protected by biome, and nitrogen use efficiency. The Organisation for Economic Co-operation and Development (OECD) has looked at good governance conditions by means of the 12 Principles on Water Governance. The results of that survey are explained in the section on Improving Water Security and Key Dimension Performance by Good Governance . With respect to KD4 the survey showed (Figure 20) that the level of dedicated water quality and preservation policies in the six regions can be improved. Even if the policies are in place, they are often only partly implemented.

Figure 20: Level of Implementation of Dedicated Water Quality and Preservation Policies per Region, 2020 (%)

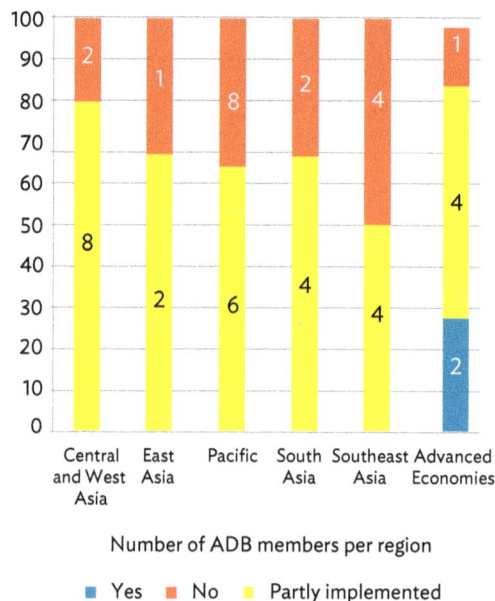

Number of ADB members per region

■ Yes   ■ No   ■ Partly implemented

Source: Asian Development Bank.

KEY DIMENSION 5
WATER-RELATED DISASTER SECURITY

KD5 assesses a nation's recent exposure to water-related disasters, their vulnerability to those disasters, and their capacity to resist and bounce back.

Indicators included in KD5

Climatological risk (drought)

Hydrological risk (flood)

Meteorological risk (storm)

KD5 direct links to the SDGs

SDG 3 Ensure healthy lives and promote well-being for all at all ages.

SDG 6 Ensure access to water and sanitation for all.

SDG 9 Build resilient infrastructure, promote sustainable industrialization, and foster innovation.

SDG 11 Make cities inclusive, safe, resilient, and sustainable.

SDG 13 Take urgent action to combat climate change and its impacts.

SDG 14 Conserve and sustainably use the oceans, seas, and marine resources.

SDG 15 Sustainably manage forests, combat desertification, halt and reverse land degradation, and halt biodiversity loss.

What do the numbers tell us about KD5?

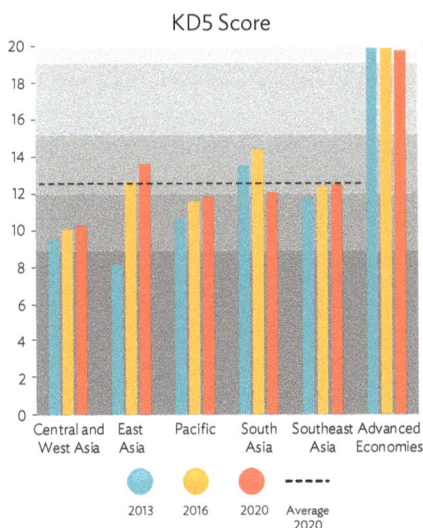

KD5 Score

Categories (right axis): 5, 4, 3, 2, 1

X-axis: Central and West Asia, East Asia, Pacific, South Asia, Southeast Asia, Advanced Economies

Legend: 2013, 2016, 2020, Average 2020

Number of ADB Members

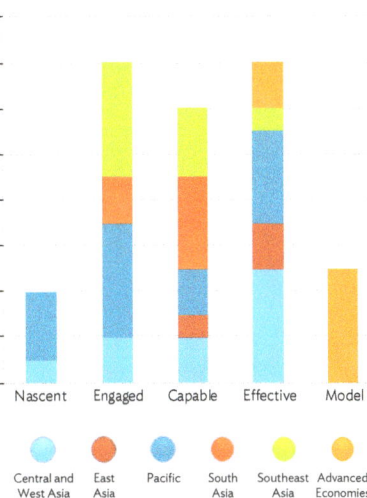

X-axis: Nascent, Engaged, Capable, Effective, Model

Legend: Central and West Asia, East Asia, Pacific, South Asia, Southeast Asia, Advanced Economies

No major differences between the regions but all of them lag way behind the Advanced Economies; Central and West Asia is lagging behind a bit but shows good progress ■

Progress is made by nearly all ADB members with the exception of some in South Asia ■

4 ADB members (Afghanistan, Kiribati, Federated States of Micronesia, and Tuvalu) are still in the nascent development stage ■

Top performers and challenged ADB members on KD5

Between 2013 and 2020 (scores up at scale of 20)

+6 points
People's Republic of China

+5 points
Kyrgyz Republic

+4 points
Niue • Taipei,China • Tajikistan

+3 points
Mongolia

Lower score in 2020

−3 points
Maldives • Tonga

−2 points
Cambodia • Fiji • India • Solomon Islands

Selected KD5-related projects

Indonesia
Flood Management in Selected River Basins Sector Project

The project will support the government and river-based communities in better managing and mitigating flood risks. The project aims to establish a GIS-based monitoring and evaluation system as well as to promote the institutionalization of a flood risk management approach.

Kyrgyz Republic
Climate Change and Disaster-Resilient Water Resources Sector Project

The project addresses flood, mudflow, and drought risks. Disaster-prone irrigation systems are prioritized, and primary and on-farm canals are modernized using an integrated and technology-supported approach.

Key Dimension 5: Water-Related Disaster Security

KD5: Water-Related Disaster Security

Water-related disaster security assesses a nation's recent exposure to water-related disasters, their vulnerability to those disasters, and their capacity to resist and bounce back.

Introduction

Achieving water security in the face of many natural hazards is a challenge all ADB members will grapple with in the coming years and decades. It is only through a concerted effort in policy, infrastructure, and disaster risk management that governments can prevent the needless loss of life and assets due to physical phenomena like floods, droughts, and cyclones. These efforts will be most effective when nations work together to share ideas and build at scale. When disaster strikes, it does not observe national boundaries. To resist its devastation, all countries must work across boundaries to build resilience. Developing alliances is essential to combating the growing threat of water-related disasters.

KD5 is a composite of three water-related disaster risk indicators:

- climatological risk (drought),
- hydrological risk (floods and mudslides), and
- meteorological risk (storms and storm surges).

KD5's definition builds on disaster risk defined by the United Nations Office for Disaster Risk

Reduction—"the potential loss of life, injury, or destroyed or damaged assets which could occur to a system, society or a community in a specific period of time, determined probabilistically as a function of hazard, exposure, vulnerability and capacity."

The hazard and exposure components of KD5 are based on disaster impact data over the last 10 years, a representative period, rather than probabilistic modeling, to align with the other AWDO KDs and allow high levels of transparency and replicability in the assessment framework. With this framework, KD5's assessment of water-related disaster security considers disasters that are climatological, hydrological, and meteorological and are quantified by their impacts on humans. However, quantifying hazard and exposure is not easy (Box 7).

Despite the connection between disasters and development, the Millennium Development Goals agenda in 2015 (before SDGs) did not focus on disaster risk and resilience. But even the present SDGs do not directly address disaster risk. That disasters undermine development gains is

Box 7: Quantifying Hazard and Exposure of Water-Related Disasters

Quantifying the hazard and exposure components of water-related disasters poses a dilemma. On the one hand, water-related disasters by their very nature do not occur frequently, and a sufficiently long period must be assessed to obtain reliable average annual estimates, especially for smaller countries. On the other hand, many countries (notably the People's Republic of China and Bangladesh) have invested in policy reform and water-related disaster infrastructure over the last few decades, reducing their exposure to water-related hazards, so a long period may not represent the current state. Sensitivity testing suggests the last 10 years to be a pragmatic compromise for estimating water-related hazards exposure under the assessment framework. Particularly devastating water-related disasters over the last decade may skew results for several countries, e.g., the 2011 floods in Thailand and the 2014–2016 droughts in the south central coastal and highland regions in Viet Nam.

Source: Asian Development Bank.

universally accepted, but the role of development approaches in exacerbating vulnerability to climate change is hardly acknowledged. Using integrated approaches—such as prevention, preparedness, and early warning systems—helps reduce disaster risks and protect the population and economy from severe weather events such as droughts, extreme rainfall, heat and cold waves, floods, and storms.[28]

This section summarizes the results of the KD5 assessment. For more details, please see the full KD5 report.[29] Appendix 7 lists the scoring approach of KD5 and the KD5 scores of the three indicators for all 49 ADB members.

Key Dimension 5 Results across the Regions

Figure 21 shows the KD5 scores (water-related disaster risk) of ADB regions. Low values indicate high water-related disaster risk. The risk in Advanced Economies is significantly lower than in all other regions. While the drivers of water-related disasters (such as hazard risk) may vary between regions, all regions are at similar and significant risk.

The KD5 scores of all five ADB regions range from 10.2 to 13.6. By looking at the disaster type scores, one can discern clear differences between each region's risks. A clear trend is that outside Advanced Economies, KD5 scores

Figure 21: Population-Weighted Average Key Dimension 5 Results of ADB Regions, 2020

	Climate Risk (Drought)	Hydrological Risk (Flood)	Meteorological Risk (Storm)	Score
Central and West Asia	2.3	2.4	5.5	10.2
East Asia	4.7	2.9	6.0	13.6
Pacific	3.5	3.7	4.6	11.8
South Asia	2.8	3.6	5.6	12.0
Southeast Asia	4.5	3.2	4.8	12.4
Advanced Economies	6.6	6.4	6.7	19.7
Asia and the Pacific*	3.7	3.2	5.6	12.5

* Without Advanced Economies.

Notes: Maximum score for KD5 is 20; numbers may not sum precisely because of rounding.

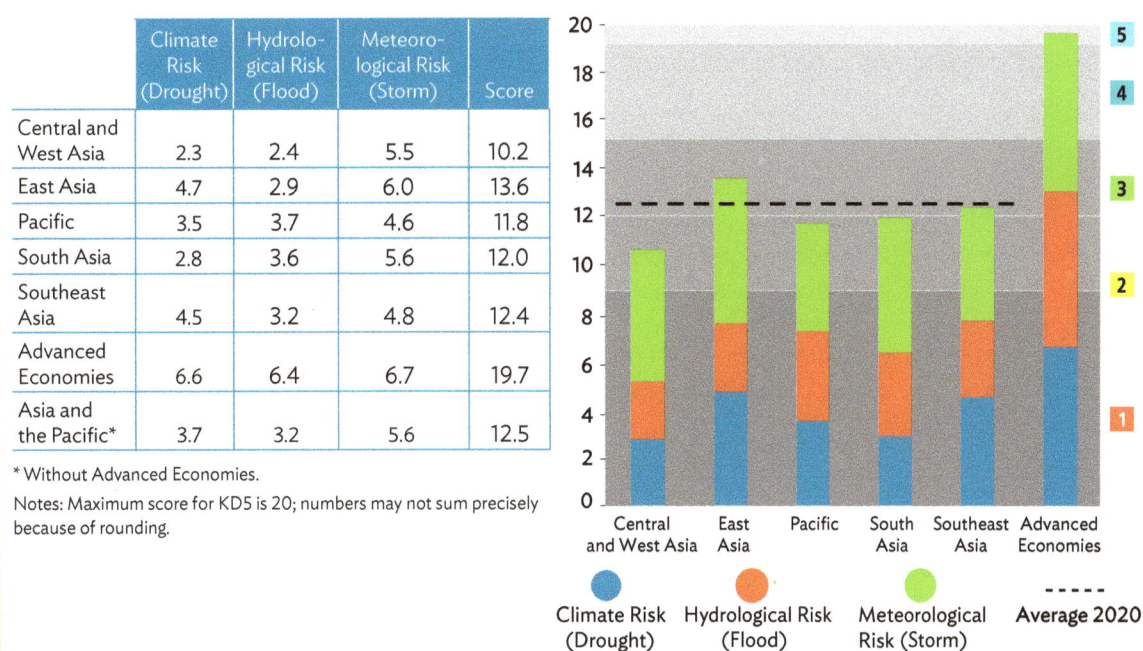

Source: Asian Development Bank.

[28] United Nations International Strategy for Disaster Reduction and World Meteorological Organization. 2012. *Disaster Risk and Resilience.* Thematic Think Piece. UN System Task Team on the Post-2015 UN Development Agenda.

[29] ADB. Forthcoming. KD5 Water-Related Disaster Security – Final Report. Korea Institute of Civil Engineering and Construction Technology – JHSUSTAIN.

for hydrological and climatological risks are lower than for meteorological risk. This trend implies that addressing poor flood and drought security is necessary to address water-related disaster risks in all regions across Asia and the Pacific.

Risks to Future Water-Related Disaster Security

KD5 measures the current level of risk. Risk is the product of three components that can change over time: hazard exposure, vulnerability, and coping capacity. Hazard exposure is exacerbated by climate change, which brings about changes to regional and local climatology and hydrology. Although climate change effects differ regionally, most of Asia and the Pacific will experience profound impacts, especially India, which has made progress in KD5 since 2010 (Box 8).

Climate change threatens recent advances in economic development and water security in Asia and the Pacific and, thus, the livelihoods of millions of people. Higher temperatures, changing rainfall patterns, more intense tropical cyclones, and sea level rise are likely to disrupt hydrological, ecological, and social systems, which imperil energy, water, and food security and aggravate existing vulnerabilities in energy supply, agriculture, and social structures.

The risk-based nature of KD5 requires activities under this dimension to pay attention to climate change. As KD5 is already oriented at dealing with climate variability, activities should be prepared for a wider range of climate variability—longer and more severe droughts, greater floods, higher sea levels, and more powerful tropical cyclones. Disregarding climate change results in reduced security against water-related disasters. Water-related disaster reduction must also be taken into account during COVID-19 (Box 9).

Box 8: Key Dimension 5 in India

Since 2010, India has made significant progress in proactive governance in disaster risk reduction. The relatively low score of key dimension 5 (KD5) (water-related disaster security) for 2020 is mainly caused by the major drought in 2015–2016, affecting 330 million people, and the increased levels of unsustainable groundwater extraction. But flood resilience has improved considerably by governmental disaster risk reduction actions, following the Hyogo Framework, alongside enhanced environmental protection.

Source: Asian Development Bank.

Box 9: Water-Related Disaster Risk Reduction in the Time of COVID-19

Water-related disasters loom large in countries under the coronavirus disease 2019 (COVID-19) pandemic. Competition and complications among disaster risk reduction emergency responses and COVID-19 health-care responses could magnify negative impacts. The High-Level Experts and Leaders Panel on Water and Disasters (HELP) developed 10 principles to address water-related disaster risk reduction under the COVID-19 pandemic. The principles offer practical advice to formulate strategies and actions. Implementation of disaster risk reduction strategies and preemptive measures that factor in the current pandemic is needed to protect areas impacted by water-related disasters from also becoming new epicenters or clusters of the pandemic. While the principles address water-related disasters, they also apply to other types of disasters.

Source: Asian Development Bank.

Water-Related Disaster Security and Gross National Income

Figure 22 shows a strong correlation as well as some variations between the KD5 scores of ADB members (at a scale of 1–20) and their GNI per capita, with the GNI plotted on a logarithmic scale. Tajikistan, the Kyrgyz Republic, Mongolia, and Georgia are the outliers on the positive side (strong security). In contrast, Thailand, the Lao PDR, Pakistan, and Cambodia exhibit weak water-related disaster security relative to their GNI. Landlocked countries do not experience storm surges, as do coastal countries, and thus perform better in KD5.

Gender and Disaster Security

In 1991, Cyclone BOB 01 killed nearly 140,000 in Bangladesh, of which 90% were women.[30] The huge gender gap in terms of impact was attributed to women's lack of ability to swim and restrictions on women to leave the house. Similarly, the 2004 tsunami killed over 230,000 people, of which 80% were women.[31] Women also accounted for 61% of the 140,000 fatalities caused by Cyclone Nargis in Myanmar in 2008.[32] While gender-disaggregated data are limited, even for major disasters, the trend is clear—women and vulnerable people are disproportionately affected by water-related

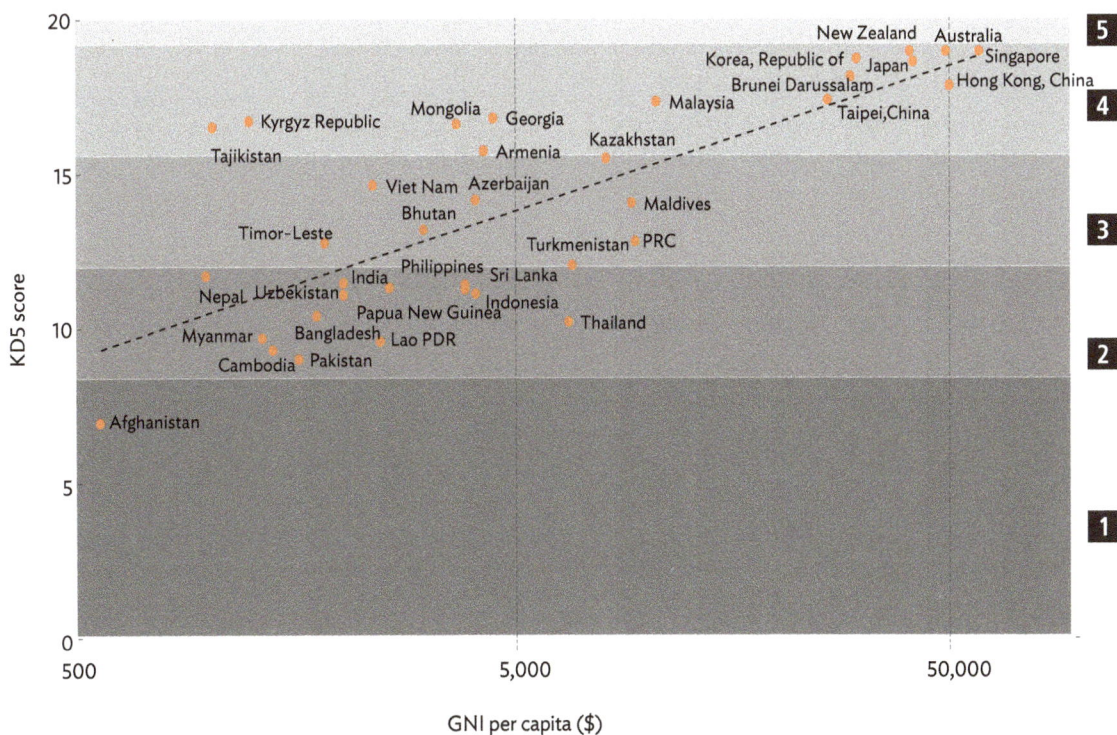

Figure 22: Water-Related Disaster Security and Gross National Income per Capita

GNI = gross national income, KD5 = key dimension 5 (water-related disaster security), Lao PDR = Lao People's Democratic Republic, PRC = People's Republic of China.

Note: Small Pacific island states are not included.

Source: Asian Development Bank.

30   Ikeda, K. 1995. Gender Differences in Human Loss and Vulnerability in Natural Disasters: A Case Study from Bangladesh. *Indian Journal of Gender Studies*. 2 (2). pp. 171–193.

31   Oxfam International. 2005. The Tsunami's Impact on Women. *Oxfam Briefing Note*.

32   Habtezion, S. 2013. Gender and Disaster Risk Reduction. *Gender and Climate Change Asia and the Pacific Policy Brief*. No. 3. New York: United Nations Development Programme.

disasters. The reasons for this disproportion vary depending on culture and the nature of disasters but generally center around the pervasive division of societies by gender that limits the social and economic resources available to women. Although more expansive data are needed, the gender gap shows that a gender approach to disaster risk management will offer the best opportunity to enhance security against water-related disasters.

Governance and Finance

The first condition for KD5 is having a dedicated water-related disaster policy. Figure 23 shows that 39 of 48 ADB members have indeed such policies, but the majority of these are only partly implemented.

KD5's second condition is financial. Water-related disaster infrastructure is capital intensive, with finance needed to cover upfront construction costs and ongoing maintenance typically repaid over long periods. Investment in water-related disaster security reduces damages, benefiting both the public and private sectors. However, many of these benefits cannot be easily monetized, undermining project viability. Investment flowing toward water-related disaster infrastructure is insufficient to bridge the gap between the investment and the requirement. For flooding, the financing gap between future needs and current investment is around $61 billion, or 0.24% of gross domestic product (GDP) of developing Asia (see section on Financing Needs for Improving Key Dimension Performance). Due to climate change, further investments in water-related disaster risk prevention and protection will be required to maintain current security levels. Public finance for water-related disaster infrastructure should increase, and governments should leverage alternative financing sources by crowding in commercial finance, climate adaptation financing, and blended financing by mobilizing private sector financial resources.

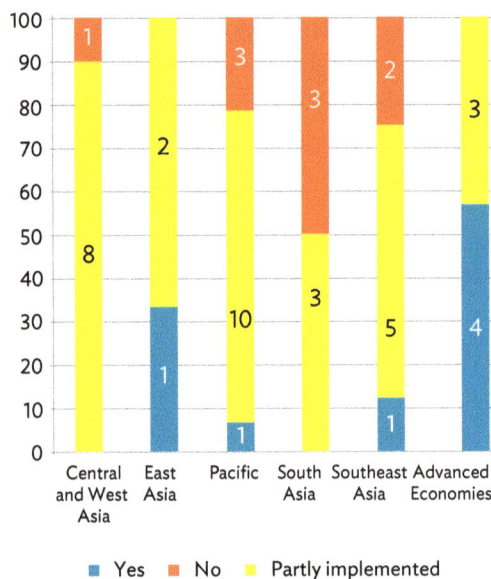

Figure 23: Level of Implementation of Dedicated Water-Related Disaster Policies per Region, 2020
(%)

Note: Numbers in the bars are the number of ADB members in the KD score.
Source: Asian Development Bank.

Integrated flood risk management—including flood risk mapping, land use planning guidelines (restrictions on land development in flood plains), and early warning systems—is a cost-effective investment to limit the exposure of people and assets to water-related disaster risks. An integrated solution, combining structural approaches and nature-based solutions with early warning systems and residual risk instruments, is crucial (Box 10).

Water-related disaster risk mitigation projects must align with national objectives and be prepared with a sound financing strategy. Developing ADB members need to address opportunities for investments in water-related risk projects by integrating them into national development planning and formulating sectoral long-term integrated plans that effectively secure commitment to investment.

Box 10: Nature-Based Solutions

Nature-based solutions can be a cost-efficient complement to gray infrastructure, contributing to building resilience to water-related disasters and providing environmental, social, and economic co-benefits. Nature-based solutions may play a significant role in stimulating economic growth in the wake of the coronavirus disease 2019 (COVID-19) pandemic and allow nations to "build back better." Examples of nature-based solutions in this context might be (i) payment for ecosystem services to farmers in exchange for the protection of catchments, and (ii) sustainable urban drainage systems.

With Asian Development Bank (ADB) assistance, some Vietnamese cities, including Vinh Yen, Hue, Ha Giang, and Ho Chi Minh City, will integrate nature-based solutions by rehabilitating their ponds, parks, and rivers, thereby greatly increasing their sustainability and climate resilience.

Source: ADB.

Improving Water Security and Key Dimension Performance by Good Governance

Achieving water security for all ADB members is a challenge due to increased water demand for productive uses such as agriculture and industry, as well as rising urbanization and per capita domestic use. The present situation for the five key dimensions is described in previous sections. Economic growth and climate change will exacerbate these challenges. As mentioned in the OECD Principles on Water Governance, addressing these issues requires coordination across all levels of government, "robust public policies, targeting measurable objectives in pre-determined time-schedules at the appropriate scale, relying on a clear assignment of duties across responsible authorities and subject to regular monitoring and evaluation."[33] The government, public and private sectors, and other stakeholders can work together to incorporate water governance into the design and implementation of such public policies.

In the AWDO 2020 framework, the OECD has surveyed 48 ADB members[34] in Asia and the Pacific using the 12 Principles on Water Governance to provide a snapshot of governance gaps in the region. These principles (Figure 24) are based on three mutually reinforcing and complementary dimensions of water governance:

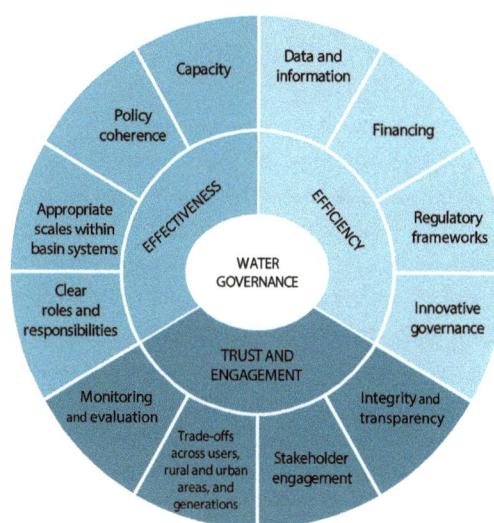

Figure 24: Water Governance Principles of the OECD

Source: Organisation for Economic Co-operation and Development (OECD). 2015. *OECD Principles on Water Governance.*

---

[33]   OECD. 2015. *OECD Principles on Water Governance.* p. 3.

[34]   Niue has recently become an ADB member and could not be included in the survey yet.

- Effectiveness of water governance
- Efficiency of water governance
- Trust and engagement in water governance

Figure 25 illustrates the general picture of the implementation of OECD principles and the regional differences in Asia and the Pacific. A high regional score expressed in green means that the principle is in place in all or most of ADB members in the region. A low regional score expressed in red means that the principle is not in place in all or most of ADB members in the region. The scores in between expressed in orange indicate that some ADB members have not implemented the principle or that the principle is partly not in place in some ADB members. Principle 8 (innovative governance) could not be appraised because of lack of data. The full results by country are shown in OECD (2020).[35]

ADB members perform rather well against principles 1 and 2, while the main governance gaps are related to principles 3–12. A more granular analysis of key findings against all principles is provided as follows.

Principle 1: Roles and responsibilities—An overarching water policy framework is in place in most ADB members.

Environmental law is in place in all 48 ADB members and water law in 75% of them. This legal framework sets the overarching principles for water policy-making in each country. For instance, the framework can support overall regulations consistently across the territory, set public service obligations, or define key water management principles. With regard to access to water and sanitation, the overarching national framework (water law or the constitution) commonly mentions the human right to water and sanitation, as recognized by the United Nations General Assembly on 28 July 2010.[36]

Principle 2: Management at the appropriate scale—Some water policy coordination mechanisms are in place in most ADB members.

The governance survey shows that 81% of ADB members have river basin organizations in place. These organizations are important institutional mechanisms for coordinating water policies at the territorial level, across stakeholders, and between government levels. They can be instrumental for integrated water management at the appropriate basin or watershed scale to reflect local conditions and foster multilevel cooperation, encourage sound hydrological cycle management, and promote climate adaptation and mitigation strategies. As such, they can help manage water risks, thus reinforcing water security.

Key Governance Gaps and Regional Priorities

Principle 3: Policy coherence—Cross-sectoral coordination mechanisms are in place, but implementing water-related policies is limited due to human resources and funding gaps.

Most ADB members have adopted dedicated water policies, with water-related disaster policies (KD5) being more widely adopted (79% of ADB members) than water, sanitation, and hygiene (WASH) policies (KD1 and KD3) (65% of the ADB members) or water quality and preservation policies (KD4) (58% of ADB members). In general, these dedicated water policies tend to clearly indicate the goals and duties of the involved water institutions. However, these policies do not clearly indicate the resources needed to achieve the goals in most ADB members, thus generating unfunded mandates and hampering their implementation. In many ADB members, the lack of financial resources (Principle 6) is compounded by a lack

35   OECD and ADB. Forthcoming. Water Governance in Asia-Pacific.
36   UN General Assembly, Resolution 64/292, The Human Right to Water and Sanitation. A/RES/64/292 (28 July 2010).

Figure 25: Regional Results Survey on Water Governance Principles of the OECD

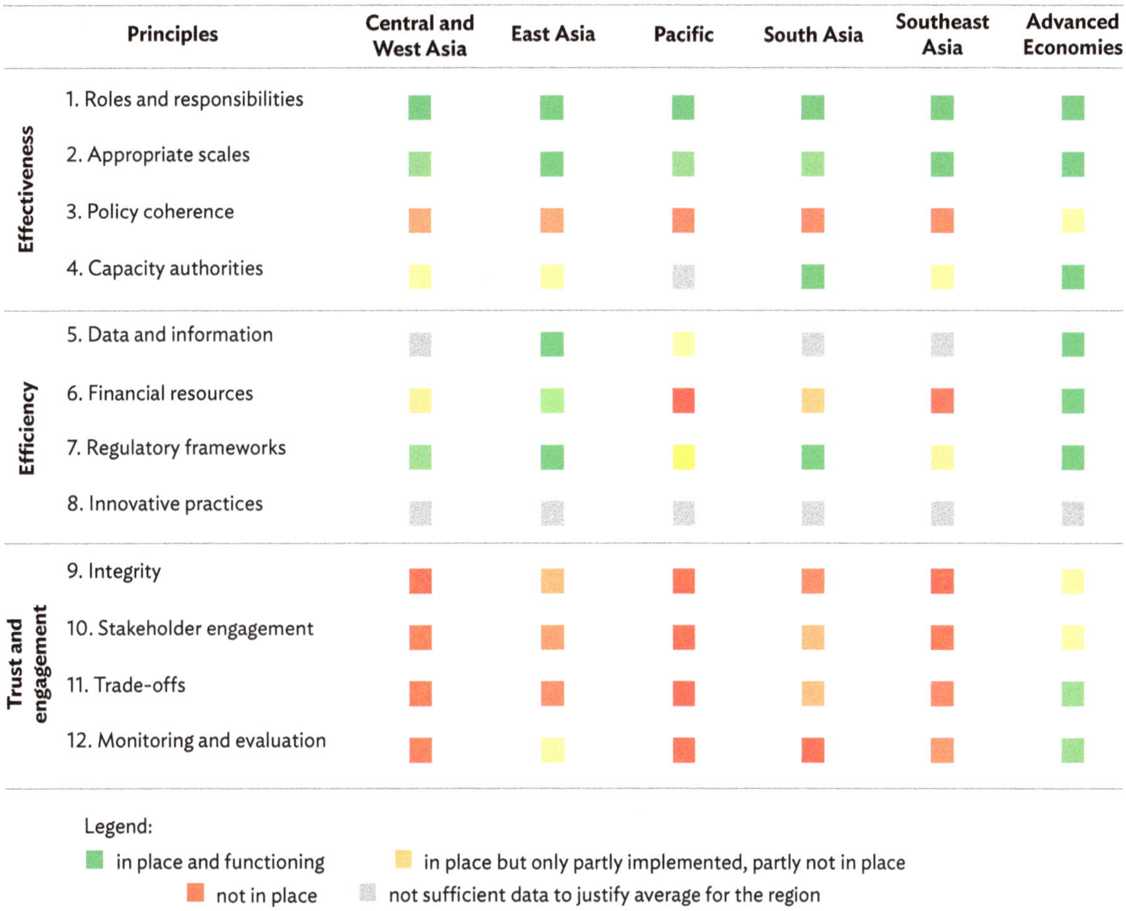

	Principles	Central and West Asia	East Asia	Pacific	South Asia	Southeast Asia	Advanced Economies
Effectiveness	1. Roles and responsibilities	🟩	🟩	🟩	🟩	🟩	🟩
	2. Appropriate scales	🟩	🟩	🟩	🟩	🟩	🟩
	3. Policy coherence	🟧	🟧	🟧	🟧	🟧	🟨
	4. Capacity authorities	🟨	🟨	⬜	🟩	🟨	🟩
Efficiency	5. Data and information	⬜	🟩	🟨	⬜	⬜	🟩
	6. Financial resources	🟨	🟩	🟥	🟧	🟥	🟩
	7. Regulatory frameworks	🟩	🟩	🟨	🟩	🟨	🟩
	8. Innovative practices	⬜	⬜	⬜	⬜	⬜	⬜
Trust and engagement	9. Integrity	🟥	🟧	🟥	🟥	🟥	🟨
	10. Stakeholder engagement	🟥	🟥	🟧	🟧	🟥	🟨
	11. Trade-offs	🟥	🟥	🟥	🟧	🟥	🟩
	12. Monitoring and evaluation	🟥	🟨	🟥	🟥	🟥	🟩

Legend:

🟩 in place and functioning   🟨 in place but only partly implemented, partly not in place

🟧 not in place   ⬜ not sufficient data to justify average for the region

Sources: Asian Development Bank and Organisation for Economic Co-operation and Development. 2015. *OECD Principles on Water Governance.*

of human resources (Principle 4), preventing the timely and efficient implementation of investment projects and dedicated water policies due to an absence of skilled staff and expertise. With regard to coordination across water-related policies and key related areas (such as health, energy, agriculture, land use), the governance survey shows that 79% of ADB members have set up coordination mechanisms at national and/or subnational levels in the form of cross-sectoral groups, meetings, reviews, research programs, etc. The majority of ADB members lacking these horizontal mechanisms

are in the Pacific (the Cook Islands, the Federated States of Micronesia, Papua New Guinea, Tonga, and Tuvalu) and Southeast Asia (Cambodia, the Lao People's Democratic Republic, and Malaysia).

Principle 4: Capacity—An important capacity gap is observed.

This lack of capacity refers not only to the technical knowledge and expertise (in terms of planning, rulemaking, project management, finance, budgeting, enforcement, risk management, and evaluation) but also to the lack of staff and time, as well as obsolete infrastructure. Specifically, it includes low capacity in public procurement, tendering, and contract management processes, and the lack of skilled staff to design, manage, and implement investment projects through the entire value chain. Therefore, it can result in a low level of capital expenditure deriving from a low absorption rate of international grants and transfers due to a low capacity impeding the implementation of investment projects. The capacity gap observed is not restricted to the subnational level, as only one-third of ADB members in Asia and the Pacific have adopted guidelines or standards for capacity building across authorities at all levels, mainly those in Advanced Economies (Australia, Japan, New Zealand, the Republic of Korea) and the Pacific (Fiji, Kiribati, the Marshall Islands, Nauru, Samoa, Solomon Islands, Vanuatu).

Principle 5: Data and information—Data are patchy and insufficient.

In many ADB members, limited information and monitoring are exacerbated by the lack of capacity, resources, and expertise to collect, analyze, and interpret water data. The governance survey reveals the widespread absence of water-related data and information across a vast majority of ADB members, except the Advanced Economies. This lack of data affects water and sanitation services, water resources management, and water-related disaster information systems. In addition, the governance survey results show that a large share of information could not be found ("No data available"), confirming that water-related data are either missing or not readily available to the public. This issue highlights the crucial need to enhance water-related data production. Missing and patchy data remain a prominent obstacle to effective water policy implementation in most ADB members surveyed, limiting the possibility for water policy monitoring and evaluation as specified in Principle 12. The KD analyses described in previous sections also underscore the lack of good information (including technical issues) for decision-making.

Principle 6: Financial resources—Notably, the use of economic instruments to manage water resources is limited.

In addition to policy instruments, economic instruments can also play a critical role in managing water risks at least cost for the community. The governance survey shows that abstraction and pollution charges are only collected in approximately one-third of ADB members in Asia and the Pacific, with more abstraction than pollution charges being implemented in many ADB members. The Pacific ADB members are where abstraction and pollution charges are the most absent, followed by Southeast Asia and, to a lesser extent, South Asia. The absence or low enforcement of robust economic instruments to manage water resources threatens the region's water security. Indeed, many ADB members are deprived of financial tools that can incentivize water users and polluters to (i) internalize the economic consequences of their water abstraction and pollution and encourage a behavioral change, and (ii) fund the costs of managing water resources and regulating activities that have an impact on water availability and quality.

Principle 7: Regulatory frameworks—Notably, the effectiveness of the water services regulation is limited.

Dedicated water services regulatory bodies have been set up in nearly all ADB members of the region, but no information could be found for one-third of the surveyed ADB members regarding the precise definition of their mandate and powers in existing bylaws, more specifically in the Pacific (the Cook Islands, Fiji, the Marshall Islands, Palau, Nauru, Tonga, Tuvalu). The regulator's success in undertaking its functions depends on the breadth and depth of the powers granted by legislation and other regulations. This unclear situation may create overlaps, competition, or conflicting objectives between regulatory bodies, line ministries, and other water-related institutions, jeopardizing the effectiveness of water policy implementation and outcomes.

Principle 9: Integrity—Uptake of integrity practices and tools is limited.

Mainstreaming integrity and transparency practices across water policies, water institutions, and water governance frameworks are key for greater accountability and trust in decision-making, and effective implementation of water policies. The governance survey underlines the low level of implementation of integrity and transparency tools. Less than 20% of ADB members (Armenia; Australia; Hong Kong, China; Japan; New Zealand; Singapore; Bhutan; India; Indonesia; Thailand; Taipei,China; and Vanuatu) have implemented relevant international conventions, or institutional anti-corruption plans or tools to track budget transparency. This low level of adoption of integrity and transparency tools can threaten water security, as investments can be discouraged by widespread corruption practices, despite considerable needs.

Principle 10: Stakeholder engagement— Mechanisms for stakeholder engagement are limited.

The accountability gap is also hampering stakeholder engagement, which is generally low in Asia and the Pacific. For instance, less than 10% of ADB members (mainly Advanced Economies) have carried out a stakeholder mapping to understand who does what in water resources and services management. This mapping can be considered as a first step to guide and build stakeholder engagement processes. Furthermore, only one-third of the surveyed ADB members have implemented formal or informal mechanisms to engage in water-related topics with stakeholders. The governance survey shows that only 27% of these ADB members have set up peer-to-peer dialogue platforms across river basin organizations or networks of river basin organizations. As such, ADB members are not reaping the full benefits of setting up river basin organizations to promote further stakeholder engagement.

Principle 11: Trade-offs—Uptake of water policy instruments to manage trade-offs is limited.

The trade-off principle is strongly linked with Principle 7 on regulatory frameworks. While water coordination mechanisms are in place in most ADB members, significant water management issues remain due to the limited adoption of water policy instruments. For instance, as the survey shows, 79% of ADB members have no policy instruments to allocate or monitor groundwater. The absence of groundwater allocation and monitoring is particularly observed in Central Asia, the Pacific, and Southeast Asia, threatening water security. Water stress will increase in Central Asia, exacerbating water crisis. In other ADB members, such as Bangladesh, northern parts of the PRC, northern India, Indonesia, and Pakistan, the lack of conjunctive management of surface and groundwater sources, together with insufficient cross-sectoral coordination between agriculture, energy, and drinking water supply also results in over-abstraction of groundwater sources. In such a context, the predominant absence of water allocation mechanisms also represents a major concern for water security in the region. The governance survey also indicates that two-thirds of ADB members have not prioritized water allocation among users in case of scarcity or emergency. This absence of allocation regime is most observed in Central and West Asia, the Pacific, East Asia, and Southeast Asia, where the risk of "too little water" is most vivid, thus jeopardizing water security further. However, despite the absence of water regimes in two-thirds of ADB members, half of them have set up mechanisms to solve water-related conflicts. It demonstrates that even if ex ante water regime policies are not commonly in place in the region, ex post mechanisms are adopted in many ADB members.

Principle 12: Monitoring and evaluation—Formal requirements for decision-making and evaluation are lacking.

Principle 5 highlights the lack of good data as a basis for decision-making and evaluation. In particular, adequate information generation and sharing among relevant actors, and scattering and fragmentation of water and environmental data are bottlenecks across ministries, agencies, and government levels involved in water policy. Moreover, even if these data were available, there are generally no formal requirements for evaluation and monitoring in two-thirds of ADB members, meaning that the implementation of dedicated water policies is hardly ever monitored. Only Solomon Islands in the Pacific; Bhutan in South Asia; Azerbaijan and Armenia in Central and West Asia; and Indonesia, Malaysia, and Viet Nam in Southeast Asia have implemented such requirements. The absence of periodical review and scrutiny of water policies prevents assessing the effectiveness of policies and potentially implementing remedial actions when policies do not deliver intended outcomes. This situation may aggravate water security further, especially when policies are being partly implemented.

Regional Priorities

Based on the above results, key areas and priority actions for governance improvement have been identified for each region. The top three priorities are policy coherence, stakeholder engagement, and trade-offs (Table 3).

Good Governance and Water Security and Key Dimension Performance

The OECD Principles on Water Governance consider good governance as a means to an end. As such,

water governance can help manage water risks and improve water security. A few observations were drawn from the governance questionnaire and the KD scores to highlight the contribution of water governance to water security in Asia and the Pacific.[37] Many ADB members displaying high scores (4 and 5) for urban water security (KD3) have adopted key performance indicators to assess and monitor water and sanitation services performance (Figure 26).

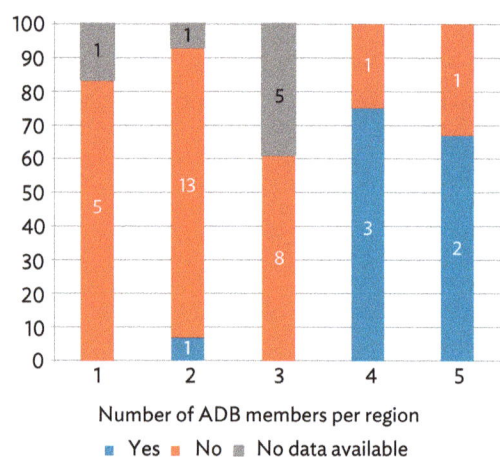

Figure 26: Share of ADB Members Adopting Key Performance Indicators for Water and Sanitation Services Based on Urban Water Security Scores

Number of ADB members per region
■ Yes   ■ No   ■ No data available

Source: Asian Development Bank.

Table 3: Priority Actions for Governance Improvement in Regions

Region	Priority Actions		
Central and West Asia	Integrity	Trade-offs	Stakeholder engagement
East Asia	Policy coherence	Trade-offs	Stakeholder engagement
Pacific	Financing	Trade-offs	Monitoring and evaluation
South Asia	Policy coherence	Trade-offs	Monitoring and evaluation
Southeast Asia	Policy coherence	Integrity	Stakeholder engagement
Advanced Economies	Policy coherence	Monitoring and evaluation	Stakeholder engagement

Source: Asian Development Bank.

---

[37]   Graphs illustrating the linkage between specific governance dimensions and KDs can be found in the appendixes.

Other observations made regarding the adoption of water policy instruments and water security are the following:

- ADB members displaying the highest urban water security (score of 5 for KD3) have all implemented both abstraction and pollution charges.
- ADB members displaying the highest water-related disaster resilience level (score of 5 for KD5) have all adopted both groundwater extractions allocation and monitoring schemes, and a water allocation regime (related to relieving droughts).
- Half of ADB members displaying the highest security score for rural household water security (score of 5 for KD1) have implemented groundwater extraction allocations and monitoring schemes.

Financing Needs for Improving Key Dimension Performance

Meeting the water-related Sustainable Development Goals (SDGs) requires improved governance (as described in the previous section); technological innovations to supply, allocate, and manage water; and a substantial and sustained financial commitment to address current and emerging challenges and provide water security for all. This section argues that improving water security in Asia and the Pacific requires huge investments for 2021–2030. Funding these investments is an issue rising on the political agenda.

Besides social and environmental reasons to improve water security, there is also a compelling economic case for water investments. Water risks must be assessed and controlled to lessen economic impacts. For example, during 2003–2013, weather-related

disasters have amounted to $750 billion losses in the region, with Myanmar, the Philippines, Bangladesh, Viet Nam, and Thailand among the most affected.[38] Water management is key for water security, climate resilience, and economic growth. A sustainable and inclusive economic growth in Asia and the Pacific requires managing water resources and mitigating water risk.

Data limitations allowed for only a partial analysis of the investment needs and financing capacities to improve water security in ADB members. The analysis carried out for AWDO 2020 covers the following three subsectors:

- water supply and sanitation (in KD1 and KD3),
- irrigation infrastructure (in one of the three indicators of KD2), and
- flood protection from rivers and seas (in two of the three indicators of KD5).

This section summarizes the results as carried out by the OECD for AWDO 2020. The full results will be described in the forthcoming report on Financing Water Security for Sustainable Growth in the Asia-Pacific Region.[39] While the KD sections address the present situation, the projected investment needs described in this section are for 2015–2030.

Investment Needs: Water Supply and Sanitation

The scores of KD1 (rural household water security) and KD3 (urban water security) show a considerable variation across ADB members in access levels to safely managed water supply and sanitation services. This variation is reflected in investment needs for water supply and sanitation infrastructure, which are driven by the following requirements:

- serve the increased number of people (particularly in urban areas),
- maintain and replace aging infrastructure assets,

---

[38] United Nations Environment Programme (UNEP). 2015. *Aligning the Financial Systems in the Asia Pacific Region to Sustainable Development*. UNEP Inquiry: Design of a Sustainable Financial System. Geneva.

[39] OECD and ADB. Forthcoming. Financing Water Security for Sustainable Growth in the Asia-Pacific Region.

- comply with increasingly stringent national and local regulations, and
- adapt to climate change.

The total annual estimated investment required over 2015–2030 to achieve universal access to safely managed water supply and sanitation services in Asia and the Pacific amounts to $198 billion per year. This estimate, which includes capital, maintenance, and operation costs, is based on World Bank figures derived from "the gap in access to services as of 2015 and the cost of connecting those without access, as well as improving the level of service for those with access to reach SDG 6.1 and 6.2 targets."[40]

Figure 27 presents the projected annual investment needs of ADB members for 2015–2030 to achieve universal access to safe water and sanitation. The PRC ($60 billion per year) and India ($22 billion per year) have the highest annual investment needs for water supply and sanitation due to the sheer size of their populations. The figure also illustrates that, except for a few notable outliers (Timor-Leste, Afghanistan, Nepal, Pakistan), most ADB members have to allocate 1%–2% of GDP to invest in water supply and sanitation infrastructure during 2015–2030, based on growth forecasts.[40]

Achieving universal access to safe water supply and sanitation services requires much more than a one-

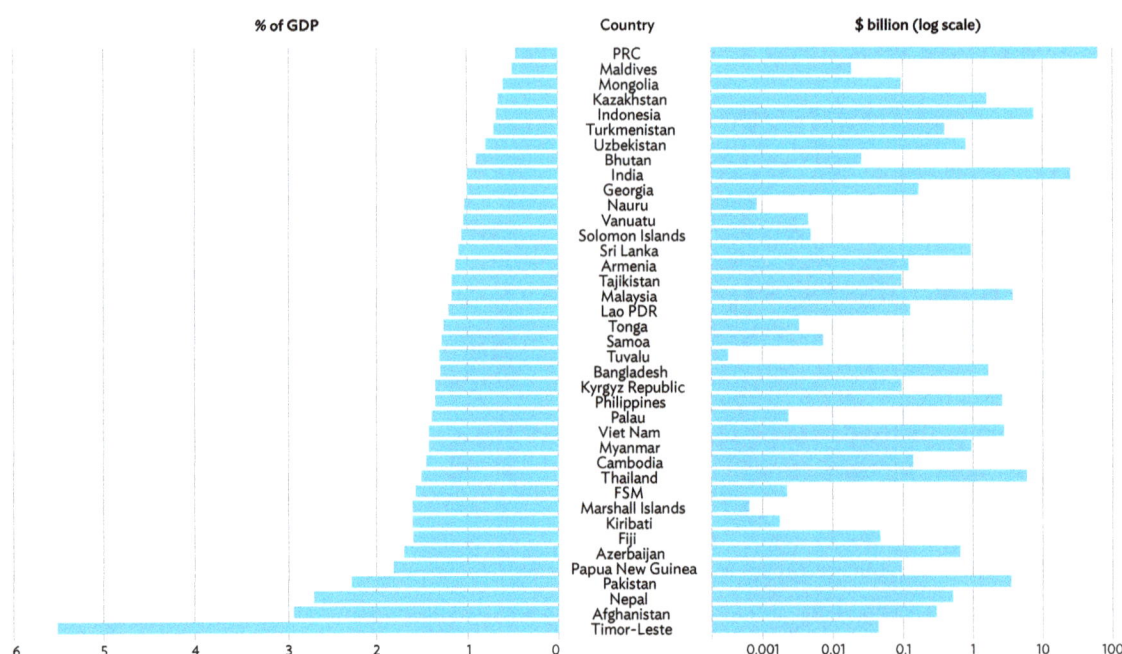

Figure 27: Projected Annual Investment Needs for Water Supply and Sanitation of Selected ADB Members, 2015–2030

FSM = Federated States of Micronesia, GDP = gross domestic product, Lao PDR = Lao People's Democratic Republic, PRC = People's Republic of China.

Source: Organisation for Economic Co-operation and Development. 2019. *Estimating Investment Needs and Financing Capacities for Water Supply and Sanitation in Asia-Pacific.* Background paper for the Roundtable on Financing Water 5th Meeting. 26–27 November.

---

[40] OECD. 2019. *Estimating Investment Needs and Financing Capacities for Water Supply and Sanitation in Asia-Pacific.* Background paper for the Roundtable on Financing Water 5th Meeting. 26–27 November. p. 4.

off injection of capital. Operations and maintenance of existing and newly built assets represent a significant share of total expenditure needs.

Investment Needs: Flood Risk Exposure

The risk of flooding is addressed in KD5 of AWDO and includes river and coastal flooding (hydrological risk). The key drivers for investing in protection against riverine and coastal flooding are climate change and socioeconomic development. These drivers are projected on three variables: the value of assets at risk of flooding, the number of people affected by floods, and the value of GDP affected by floods. The impacts on people and the scale of investment needs in flood protection (like water supply and sanitation) are mostly concentrated in low- and middle-income ADB members. Bangladesh, Myanmar, Viet Nam, and Cambodia have the greatest percentage of the population exposed to flood risks. Bangladesh, in particular, is

a hotspot for flood risk in Asia and the Pacific, with over 11% of the population projected to be exposed by 2030.

Figure 28 presents the projected yearly flood risk exposure developed separately from KD5 through probabilistic modeling. It shows that Bangladesh, Cambodia, Afghanistan, the Kyrgyz Republic, Tajikistan, and Viet Nam all have flood risks exceeding 6% of GDP by 2030 under a business-as-usual scenario, with land subsidence. Coastal flood risks are projected to strongly affect Solomon Islands, Bangladesh, Vanuatu, and Viet Nam.

How exposure and vulnerability to flood risk translate into investment needs depends on two variables not documented in this report—(i) the current level of expenditures for flood protection (very few countries globally monitor and report it accurately), and (ii) the measures to address flood risks at present and in the future.

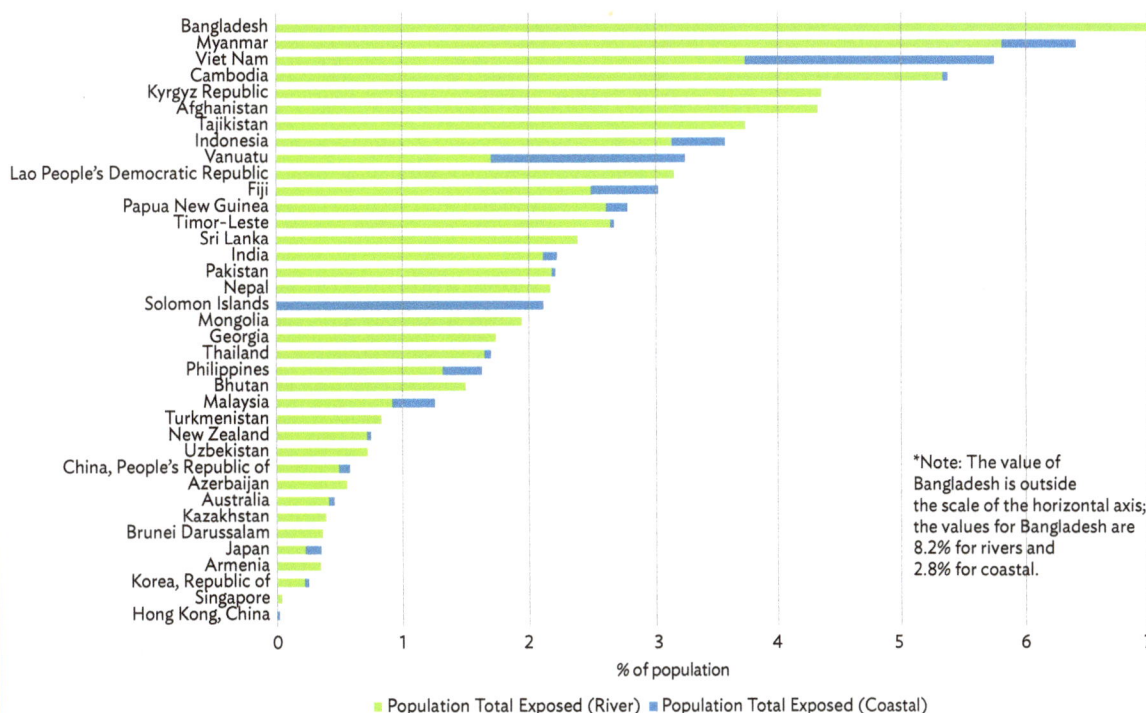

Figure 28: Project Flood Risk Exposure in Selected ADB Members, 2030
(% of population)

*Note: The value of Bangladesh is outside the scale of the horizontal axis; the values for Bangladesh are 8.2% for rivers and 2.8% for coastal.

Source: P. Ward et al. 2020. Aqueduct Floods Methodology. Technical Note. Washington, DC: World Resources Institute.

The investment level for protection against flood risks will depend primarily on the level of risk acceptable to local populations, measures taken to minimize exposure and vulnerability, and the uncertainty of construction costs. Determining that investment level is a political process in which the number of people, the value of assets affected, and the impacts of flooding on economic production are the main discussion points, combined with the financial ability of ADB members.

Investment Needs: Irrigation

Water security for agriculture, as addressed in KD2, mainly relates to irrigation. As food demand increases from a growing population, dietary preferences change, and climate change effects intensify, so too will the need for irrigation, both in terms of expansion and efficiency improvements of existing irrigation infrastructure.[41] The largest expansion in irrigation is projected to be in Asia. At the same time, urbanization will reduce the available land due to land conversion. Approximately 2.6 million square kilometers of agricultural land in Asia is irrigated (about 70% of the world's total irrigated land), of which the largest shares are in South Asia (India and Pakistan) and East Asia (the PRC).[42] "Existing projections suggest that the area under irrigation is set to expand by an additional 57 million hectares [570,000 square kilometers] by 2050, a 20% expansion from 2010 levels."[43] Irrigation expansion is projected to be particularly high in South Asia—up to a 30% increase from 2010 to 2050.

Irrigation expansion requires investments in water infrastructure such as irrigation technologies, dams, canals, and other conveyance systems.[44] Country-level data on investment needs or current expenditure on irrigation are not available. At the regional level (East Asia, the Pacific, and South Asia), the investments required to achieve projected irrigation expansion are estimated to cost on average a total of $3.1 billion per year during 2015–2030.[45]

However, irrigation expansion alone is not enough, with water scarcity leading to declining average yields, particularly in South Asia. Most of the gains from irrigation expansion are only realized if they are accompanied by investments to modernize systems and increase water-use efficiency of existing irrigation assets. Projected investments in improved water use efficiency across developing ADB members in East Asia, the Pacific, and South Asia are a significant share ($1.7 billion per year) of the cost, given the large share of land under irrigation. Baseline investments in soil and water management technologies (such as no-till agriculture and water harvesting that increase the water holding capacity of soil) are estimated at $500 million per year across the regions.

An alternative scenario combines accelerating irrigation expansion and further improving irrigation efficiency and soil and water management, increasing agricultural output while conserving more water. Under this scenario, the total annual investment in irrigation for East Asia and the Pacific is $6.8 billion per year and for South Asia is $5.1 billion per year.

Another study estimates that South Asian countries will bear the highest annual investment cost of 0.27% of their GDP, compared with 0.13% for East Asia and the Pacific and 0.04% for Central Asia.[46]

---

[41] ADB. 2017. *Financing Asian Irrigation: Choices before Us.* Manila.

[42] Asian Infrastructure Investment Bank. 2019. *Asian Water Sector Analysis: A Technical Background for the Asian Infrastructure Investment Bank (AIIB): Water Sector Strategy.* Beijing.

[43] Footnote 42, p. 18.

[44] Other needs, such as transport and logistics connecting irrigated land to markets, are not factored in. The lack of these elements can stifle irrigation expansion.

[45] Rosegrant, M. W. 2017. *Quantitative Foresight Modeling to Inform the CGIAR Research Portfolio.* Project Report. Washington, DC: International Food Policy Research Institute.

[46] Rozenberg, J. and M. Fay. 2019. *Beyond the Gap: How Countries Can Afford the Infrastructure They Need while Protecting the Planet.* Sustainable Infrastructure. Washington, DC: World Bank.

Financing Capacities to Reach Investment Needs: Water Supply and Sanitation

In some ADB members where data are available, governments and international assistance are the main sources of funding for water supply and sanitation. Households contribute significantly only in a few ADB members in Central Asia (Azerbaijan, the Kyrgyz Republic), South Asia (Bangladesh, Nepal, Pakistan), or small states (Solomon Islands), reflecting either an effort to cover costs through revenue from water bills (in Bangladesh) or the paucity of public budgets allocated to water supply and sanitation.

Water affordability is an issue in urban environments in several ADB members. Based on available data, Bangladesh, Indonesia, Mongolia, Myanmar, the Philippines, and Viet Nam face high affordability constraints, with annual tariffs in selected cities representing more than 10% of the annual income of the middle quintile household. Conversely, there may be room for maneuver to increase water supply and sanitation tariffs in some ADB members such as Armenia, Bhutan, Kazakhstan, Malaysia, and Tajikistan.

It is noted that the investment needs in big countries such as India are massive. The size of the country and the population poses not only a financial but also an organizational challenge for the government to have regional impact.

Potential Sources of Funding and Financing

The OECD distinguishes the three financial sources for water-related investments (revenues from tariffs, taxes, and transfers from the international community, aka 3Ts) and other repayable financial sources (loans, bonds, etc.). Taxes and tariffs are important not only for raising revenue but also for managing water demand and signaling water value, water services, and water security. Repayable financial sources require a creditworthy borrower, which can provide a financial return.

Notably, there is a growing consensus that mobilizing commercial finance (e.g., through blended finance or other means such as a combination of equity and sector debt) will help achieve the SDGs and provide the incentives to put the water sectors' financing on a more sustainable footing. Coordination among development finance providers will make this happen and avoid concessional finance crowding out commercial capita.[47] Examples of blended finance approaches in the region include the Philippine Water Revolving Fund; the Agence Française de Développement access to finance program in Cambodia; and microfinance interventions in Indonesia, the Philippines, and India, among others.

Country-level data describing current expenditure and financial sources for water-related investments are limited, preventing the construction of a robust and comparable expenditure baseline and the mobilization of additional finance.

Public taxes are the main financial source for water-related infrastructure. Official development assistance remains a low share of investment in water infrastructure and may not be targeting those ADB members who need it most. Further, water supply and sanitation tariffs are underutilized. While affordability acts as a barrier in some ADB members (pricing is too low to cover costs), the willingness of authorities to charge may also be low in other contexts.

Bridging the Investment Gap

The magnitude of capital investment needs and operation and maintenance costs for water supply and sanitation services, flood protection, and irrigation infrastructure calls for a shift in how the sector is currently operated, regulated, and financed. ADB members would benefit from more systematically exploring a combination of three policy options, to be tailored to national and local circumstances. These policy options are described in the next chapter of this report.

---

[47]   OECD. 2019. *Making Blended Finance Work for Water and Sanitation: Unlocking Commercial Finance for SDG6. OECD Studies on Water.* Paris.

Beneficiary of ADB's Southeast Gobi Urban and Border Town
Development Project in Mongolia

PART III

AWDO 2020 Policy Recommendations and ADB Water Strategic Directions 2030: A Water-Secure and Resilient Asia and the Pacific

Introduction

Water security is crucial for meeting the Sustainable Development Goals (SDGs), especially SDG 6, which ensures access to water and sanitation for all. It is also relevant to other SDGs addressing food and energy security, good health and well-being, sustainable cities and communities, gender equity, reduced inequalities, climate action, ecosystem health, and combating poverty. In July 2016, the United Nations (UN) launched the SDG 6 Global Acceleration Framework, an event hosted by UN-Water. In early 2020, the UN Secretary-General António Guterres called for a Decade of Action toward delivering the global SDGs.

The current coronavirus disease 2019 (COVID-19) pandemic has been costly to Asian economies[48] with predicted gross regional economic impacts estimates of 6.2% to 9.3% of regional GDP,[49] before factoring government policy responses, which could reduce these estimates by 30%–40% depending on the containment scenario. It has further reinforced the fundamental need for making water, sanitation, and hygiene available to everyone, eliminating inequalities and leaving nobody behind, especially the most vulnerable. It is a reminder of the fundamental linkages between water, health, and ecological security.

Against this backdrop, the international community—through the SDGs and other global commitments such as the Paris Climate Agreement and associated Nationally Determined Contributions, the Sendai Framework for Disaster Risk Reduction, and the Convention on Biological Diversity—has placed water security at the center of sustainable and resilient development. In addition, ADB joined world leaders and heads of international organizations with the World Health Organization's 14 May 2020 call for action on COVID-19 to prioritize and accelerate WASH access.

As such, water underpins social and economic development and is critical for realizing ADB's Strategy 2030, with aspirations aligned with major global commitments. Under Strategy 2030, ADB aims at sustaining its efforts to eradicate extreme poverty and expanding its vision to achieve a prosperous, inclusive, resilient, and sustainable Asia and the Pacific. This strategy provides a framework for the next decade through a set of operational priorities (Figure 29) and delivery mechanisms by recognizing the importance of a differentiated approach and considering societies' needs at different stages of development and varying endowments of human and natural resources.

Strategy 2030 focuses on helping the poor and disadvantaged, integrating solutions to improve the quality of life in cities and rural areas, supporting food security, recognizing the potential of gender balance in access to resources, improving governance arrangements to deliver services, building resilience and reducing exposure to disasters, safeguarding the natural environment, and encouraging cooperation beyond borders to leverage better outcomes. Integral to all operational priorities is improving water security equitably. Thus, Strategy 2030 goals and the thematic focus of its operational priorities require investments in water security and related capacity building and policies within a more integrated approach with other sectors.

A robust integration is needed to achieve Strategy 2030's vision, where the complementarity of design and inclusiveness of the process will help meet the desired regional development outcomes. This is clearly the case for operational priority 4 (making cities more livable), where water crosscuts across all three outcomes of (i) improving access, quality, and reliability of services; (ii) strengthening urban planning and financial sustainability; and (iii) improving the urban environment, climate

48 International Monetary Fund. 2020. COVID-19 Pandemic and the Asia-Pacific Region: Lowest Growth Since the 1960s. Blog. 15 April.
49 Park, C-Y. et al. 2020. An Updated Assessment of the Economic Impact of COVID-19. ADB Briefs. No. 133. Manila: ADB.

Figure 29: Strategy 2030's Seven Operational Priorities

Addressing remaining poverty and reducing inequalities

Accelerating progress in gender equality

Tackling climate change, building climate and disaster resilience, and enhancing environmental sustainability

Making cities more livable

Promoting rural development and food security

Strengthening governance and institutional capacity

Fostering regional cooperation and integration

Source: Asian Development Bank.

resilience, and disaster management. Similarly, achieving a circular economy for water, sanitation, and health requires that the supply side, service delivery, and waste reuse be considered part of a service and value chain.

In terms of water for agriculture, Strategy 2030 envisages a more integrated agenda with a renewed focus on rural development and improving market connectivity by transforming agricultural value chain links and resilient food systems (operational priority 5: promoting rural development and food security). Addressing remaining poverty and reducing inequalities (operational priority 1) also entails preventing and protecting human health and providing universal access to water, sanitation, and hygiene in an integrated and at scale manner. Associated integrated policy responses are thus fundamental to assisting Asia and the Pacific in addressing the economic impacts of global health shocks such as COVID-19. These policies should include interventions that help restructure the economy toward higher productivity growth. Investments in the delivery of health and water, sanitation, and hygiene (WASH) services can minimize disruptions to the economy and

contribute to productivity enhancement and inclusive growth strategies.

The failure to implement innovative and integrated approaches can result not only in unrealized prosperity and inclusiveness as economies develop but also in a lack of sustainability and resilience in the face of environmental, social, and economic shocks. Approaches that incorporate inter-sectoral and thematic processes can be used as tools for (i) building effective, integrated, and evidence-based design; (ii) implementing water security policy, projects, and development strategies; and (iii) filling knowledge and capacity gaps necessary to address evolving development challenges such as water security.

This need for an integrated and multisector water security assessment is a core motivator for developing the AWDO multidimensional methodology and associated regional dialogue, which ADB and the Asia-Pacific Water Forum have championed since 2007 and has been supported and enriched by additional partners such as the Australian Water Partnership, International Water Management Institute, and the OECD.

AWDO has been used to assess water security in Asia and the Pacific in 2013, 2016, and 2020. Its integrated methodological approach is completely aligned with the necessary implementation approach for Strategy 2030 to successfully address ADB developing members' key developmental challenges, including water security.

In this context, AWDO 2020 becomes a useful tool for ADB members, with its integrated methodological approach involving key dimensions (KDs) related to water security: rural (KD1), economic (KD2), urban (KD3), environmental (KD4), and water-related disaster (KD5). AWDO can be used to determine the current status and progress of KDs in relation to secure and resilient water resources management and the necessary policies toward achieving international commitments such as the SDGs and others related to climate change, disaster risk resilience, health security, and inclusive and sustainable future for all as targeted by Strategy 2030.

The interdependence of the factors that determine water security in each KD means that increases in water security will be achieved by governments that break the traditional sector silos to find the means for managing the linkages, synergies, and trade-offs among the dimensions. This process is known as integrated water resources management (IWRM), adopted by world leaders in Johannesburg in 2002 at the World Summit on Sustainable Development and reaffirmed at the UN Conference on Sustainable Development (Rio+20) in 2012. IWRM is now included in the SDGs as target 6.5.

In addition, climate change is expected to impact all KDs of water security, with changing rainfall and runoff patterns affecting water availability to meet water demand. Aquatic ecosystems and thus environmental water security (KD4) may face escalating impacts on many levels. Further, more extreme rainfall and typhoons, sea level increase, and droughts will affect water-related disaster security (KD5). Water security issues under climate change requires decision-making under a dynamic and changing future.

Achieving Water Security

AWDO KDs and water security governance and finance are important to achieve a water-secure and resilient Asia and the Pacific. Some insights can be gained by deep diving into specific policy recommendations raised by KDs as well as the governance and finance aspects of water security.

Policy Recommendations by Individual Key Dimension

Key Dimension 1: Rural Household Water Security

Increasing rural household water security is particularly relevant to Asia and the Pacific because, despite urbanization trends, nearly half of the region's households are still in rural communities. Rural households tend to be poorer and more disenfranchised than urban households. Furthermore, water and sanitation for households is generally less attractive to funding organizations, as the return on investment is lower and indirect, compared with water for economic uses like agriculture. The following policy recommendations are crucial to rural household water security.

Involve vulnerable people in decision-making. Vulnerable people refer to women, children, the elderly, the poor, people with a disability, ethnic minority, and sexual minority. Because they have a vast array of needs, they respond to policies differently. Thus, a one-size-fits-all model will not be effective. Of the 23 countries that completed the appropriate part of the 2019 Global Analysis and Assessment of Sanitation and Drinking-Water (GLAAS) survey, 19 have specific policies that include vulnerable people in decision-making. However, it is generally unknown how well-funded the needs of vulnerable people are in financially stretched WASH sectors. To achieve better outcomes for vulnerable people, governments should invest more in engaging with vulnerable groups through targeted policies and empowering them to participate in decision-making, paying special attention to women's role. Figure 30 illustrates women's role in collecting drinking water in some Asia and Pacific countries.

Figure 30: How Water Is Collected in Asia and Pacific Households
(%)

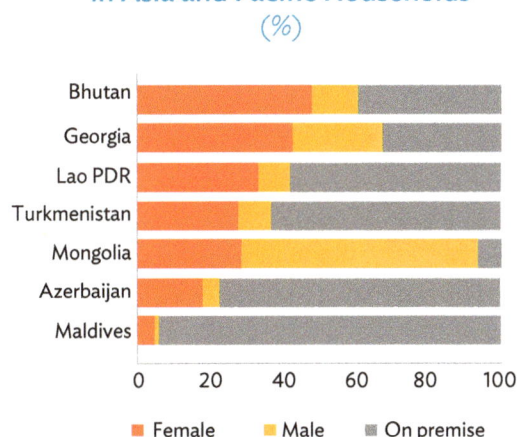

Lao PDR = Lao People's Democratic Republic.
Source: Asian Development Bank and UN Women. 2018. *Gender Equality and the Sustainable Development Goals in Asia and the Pacific: Baseline and Pathways for Transformative Change by 2030*. Bangkok.

Invest in human resources capacity. Only 2 of 31 surveyed ADB members reported that they have the human and financial resources required to implement their water and sanitation policies. This shortfall is especially concerning, as it is often poor implementation rather than poor policy that reduces access to WASH services. Without the necessary resourcing and investment in human resources, access to WASH services is likely to become a key factor limiting growth in rural WASH provision, particularly in Southeast Asia. Asian governments should immediately invest in the human resources required to meet the future's water demands.

Deliver locally appropriate solutions for Pacific nations. The population-weighted KD1 scores for Pacific ADB members are very low, and there has been relatively little progress since 2013. Due to the steady accumulation of climate change impacts, these scores may stagnate or even decrease. Papua New Guinea, for example, has shown declining rates of water and sanitation access due to increasing population and comparatively limited growth in service provision. While research into IWRM suggests that a focus on a "ridge to reef" catchment planning could be effective,[50] the human and financial resources needed to develop tailored plans are lacking. Low economic growth and geographically dispersed populations make service provision especially costly and difficult. Approaches that consider land and water management and temporal variability are needed, especially when accounting for complex and uncertain climate change impacts. Diversifying water sources available to Pacific nations is an option, e.g., rainwater tanks, which currently have a much greater uptake in rural Thailand than in rural Papua New Guinea despite similar monsoonal climates.

Key Dimension 2: Economic Water Security

Water scarcity is still increasing in Asia and the Pacific. Water withdrawals for energy and industry are expected to grow rapidly, as those sectors develop faster than agriculture. In addition, climate change will exacerbate water-related risks. Without adequate attention, declines in economic water security are possible for some ADB members. The following policy recommendations minimize the risks and are particularly relevant to economic water security.

Enhance water resources monitoring, measurement, and data availability. Optimizing water resources management requires good measurement and accurate accounting of multiple

---

50   Hadwen, W. L. et al. 2015. Putting WASH in the Water Cycle: Climate Change, Water Resources and the Future of Water, Sanitation and Hygiene Challenges in Pacific Island Countries. *Journal of Water, Sanitation and Hygiene for Development.* 5 (2). pp. 183–191.

factors, including water supply, demands, efficiencies, return flows, quality, location, uses, costs, benefits, and revenues. Gaps in data describing these factors are often cited as constraints on assessing economic water security (KD2) and improved management, so data availability has been included to indicate how governance can affect KD2. Data availability is as important as data collection and monitoring, allowing a range of users and organizations to independently examine the data and discover innovative ways to improve water management and security. In terms of monitoring and reporting, certain water balance components (e.g., evapotranspiration and groundwater storage) suffer from the largest gaps in data availability. Other monitoring-related knowledge gaps include water quality and complete records of infrastructure capital and operation and maintenance costs. Methodologically, the focus should be on ensuring that water is considered an economic input to production so that the value of water in production can be accounted for. Satellite-based information and new technologies like smart metering are continually making data collection easier, cheaper, and more possible. These technologies and improved governance of water data enhance water data availability. National policy changes might be needed to finance data archives and accessibility, ensuring regular updates, quality checks, and safe storage.

Improve water productivity. As water becomes increasingly scarce, and demands for water, food, and other goods and services increase with growing populations and incomes, water must be used more intensively while still protecting the environment and ecosystems that sustain production. Water is a continuous cycle, so water used today can be available for reuse in the future. However, water removed from water supply systems will not be available for immediate reuse, and doing so is costly. Water used more efficiently will reduce supply and treatment costs and ensure that water is available for more economic uses.

Ensure adequate storage and distribution mechanisms. Water is rarely at the right location and time and rarely in the right quality to meet any demand. Where there is a mismatch between supply and demand, water must be stored, treated, and transported. Storage is also necessary to mitigate flood and drought risks, which are projected to increase over time with climate change, thus providing opportunities for climate adaptation and resilience enhancement. Multiple options are available to increase water storage (including dams), improved land use and soil moisture management, rainwater harvesting, aquifer recharge, and river course management. Institutions should be developed to effectively monitor, maintain, enforce, and allocate water use priorities.

Promote integrated water resources management. Water is necessary for human and environmental health and economic production across all sectors. Interactions with multiple systems at any stage in the water cycle affect water availability, quality, and demand, which cannot be managed as separate entities. Economic water security requires optimizing a complex and interconnected set of systems. Institutional arrangements must account for the integrated nature of water and resource management across sectors and disciplines, which basin planning brings all together. Without good basin planning, considering water productivity improvements, allocations, and compliance monitoring is impossible.

Invest in climate change adaptation and resilient measures. Climate change may exacerbate water-related risks—but can also be used to motivate activities to improve economic water security. The threat of increased climate-related vulnerabilities and risks provides an opportunity to improve economic water security through investment in climate adaptation and resilience programs such as enhancing storage capacity and strengthening insurance systems and mechanisms. It also underscores attempts to revamp various aspects of governance, such as data availability. ADB members vulnerable to climate change impacts are recommended to devise similar strategies to improve various measures of economic water security.

KD3: Urban Water Security

Asia and the Pacific is rapidly growing and urbanizing, and Asian cities have become vital economic drivers. Water plays an essential role in achieving sustainable, livable, resilient, and productive cities. However, rapid economic and population growth and climate change create significant challenges for the provision of water, wastewater, and stormwater infrastructure. The following policy recommendations meet these challenges and are particularly relevant in urban water security.

Invest in sanitation, wastewater treatment, and the circular economy. Improvement in sanitation is needed broadly throughout Southeast Asia, Central and West Asia, South Asia, and the Pacific. Investment in wastewater collection and treatment technologies reduces pollutants and benefits water catchments and supplies. With many ADB members facing simultaneous water supply and sanitation challenges, there may be opportunities to solve both problems conjunctively by, for example, increasing investment in wastewater recycling to offset demand for potable water. Direct and indirect potable wastewater recycling technologies have been implemented elsewhere and could, if undertaken systematically and adequately managed, provide simultaneous water supply and wastewater solutions. Such solutions can be more cost-efficient (and at lower energy demand) than many traditional water supply options. Likewise, investment in improved biosolids management may help realize the value of wastewater treatment by-products, thereby supporting local economies through enhanced agricultural production while mitigating traditional waste management risks.

Foster urban water cost-effectiveness and affordability. A decrease in water affordability over time was observed for all regions except Central and West Asia. As water affordability is significantly worse and economic damage due to storms is higher in the Pacific than in other regions, water innovation in the Pacific is crucial. Affordability issues are compounded by low service coverage and high economic costs due to storms.

The relationship between energy and water security, effective asset and water demand management, and water supply energy efficiency are key cost management elements. In addition, thinking of cities as water supply catchments—and designing them for that purpose—allows a significant paradigm shift and large forward steps. Urban water security can be improved in some low-scoring nations through investment in suitable centralized and decentralized water infrastructure solutions to provide cost-effective access to reliable, diversified, and raw water sources.

Improve drainage security. Improved flooding and drainage security is broadly needed in the Pacific, Central and West Asia, and Southeast Asia. Priorities should be given to (i) a better understanding of flooding risks in and around urban areas to inform urban land use planning and investment, (ii) increased investment in catchment management to reduce watershed degradation and improve raw water quality and drinking water treatment, and (iii) enhanced flood mitigation infrastructure and operation.

Key Dimension 4: Environmental Water Security

Maintaining the health of rivers, wetlands, and groundwater systems and measuring progress on restoring aquatic ecosystems to health are vital components of water security and water use sustainability in Asia and the Pacific. The following policy recommendations ensure environmental water security.

Enhance pollution load management and circular economy. Inefficient agricultural practices concerning fertilizer use are linked to excess nitrogen and phosphorous entering nearby waterways, which can lead to algal blooms, lower dissolved oxygen, and fish kills, as well as direct harm to humans. Additionally, overreliance on fertilizers to produce greater yields can cause water quality issues in groundwater reserves and acidification problems in soils. Maximizing nutrient input to yield rates will substantially lower the rate of nitrogen entering the environment. Developing novel financial mechanisms, including nitrogen markets, is a feasible avenue to improve the sustainable use of fertilizers in agriculture and reduce nonpoint pollution loads. Point-source pollutants introduced into waterways from household and industrial sources include medical, human, and food wastes; pathogenic organisms; and poisonous chemicals such as pesticides, synthetic chemicals, and heavy metals. While the amount of wastewater is expected to increase in developing ADB members, particularly in the industrial sector, it only needs to be treated to a standard acceptable for its intended reuse (provided such water for reuse is not released back into aquatic systems). ADB members with a low wastewater score should be supported to develop localized water treatment facilities and water recovery programs that allow related or nearby businesses to operate in industrial symbiosis with regard to their water consumption. These programs will help increase wastewater treatment while reducing the amount of untreated wastewater released into natural systems.

Increase terrestrial protection. Protected areas are essential to preserving natural systems. They can have positive impacts on aquatic ecosystem health on many levels, including the (i) conservation of riparian vegetation, (ii) absence of agricultural practices, (iii) preservation of instream dynamics in the absence or minimization of water resource developments, and (iv) restrictions to fishing and other extractive industries. Of particular concern is the threat to vegetated riparian areas that filter fertilizer and excess sediment runoff from nearby agricultural operations, causing damaging instream imbalances such as toxic algal blooms. Consequently, increasing protected areas and incentivizing restoration efforts along waterways, including vegetated retention basins, and implementing regulations to impede or stop riparian vegetation removal will have direct aquatic ecosystem health benefits. Where resistance to

the introduction of protected areas is related to potential financial security impacts, economic research into possible outcome co-benefits may help break the impasse. Supporting research on the economic value of ecosystem services will further highlight potential co-benefits and encourage the implementation of protected areas.

Promote sustainable hydrological alteration and riverine connectivity. Flow alteration of river and other wetland systems is a primary cause of reduced aquatic ecosystem health, leading to reduced water quality, habitat, and biodiversity losses and facilitating the invasion of exotic species (Box 11). Maintaining adequate longitudinal and lateral riverine connectivity is particularly important for ADB members that support major freshwater fisheries with species that rely on long-distance migration for their survival. Nevertheless, flow alteration is an inevitable water resource development for urban water supply and agricultural water use. ADB members with comparatively poor hydrological alteration outcomes should be supported to develop locally specific environmental flow programs to ensure that further water resource development does not unnecessarily impair aquatic ecosystem health. This program will require multi-stakeholder engagement and substantial expert scientific advice to balance water allocation needs of communities with the requirements for positive ecological outcomes.

Address groundwater depletion. Twenty-one ADB members[51] received the lowest possible rating for groundwater resource sustainability. Overuse of groundwater leading to long-term depletion can dramatically impact surface waters and surrounding vegetation, and consequently aquatic ecosystem health and other groundwater-dependent ecosystems and stygofauna communities. ADB members with poor groundwater sustainability scores should be supported to improve groundwater use efficiency. Aquifer mapping for water

[51]   Afghanistan; Armenia; Azerbaijan; Bangladesh; Bhutan; Fiji; Hong Kong, China; India; Kazakhstan; the Kyrgyz Republic; the Marshall Islands; the Federated States of Micronesia; Nepal; Palau; Pakistan; the PRC; the Republic of Korea; Tajikistan; Thailand; Tonga; and Turkmenistan.

Box 11: Conceptual Model Showing Alterations of Human Activities Affecting River Condition and Health

The condition or health of freshwater ecosystems is defined, to a great extent, by the landscape through which it flows. Physical changes in the landscape dramatically alter the nutrients dynamics; hydrological conditions; instream morphology; habitat structure; and, subsequently, the assemblages of fish, invertebrates, plants, and algae. The catchment vegetation removal can dramatically increase the rainfall-runoff rate, altering the natural flow regime. Equally, the conversion of catchment vegetation to agricultural lands can increase the flow of nutrients to the river channel, leading to an increased risk of problematic algal blooms.

Sources: Healthy Land and Water; and Asian Development Bank (ADB). *Forthcoming. KD4 Environmental Water Security – Final Report.* Manila: ADB / International WaterCentre.

management at the aquifer or groundwater basin level can help water users coordinate and reduce their relative use to sustainable levels. Good examples in this respect are the major national aquifer mapping and management programs in India as well as the many water conservation schemes implemented by states in India. Financial incentives for water conservation (as well as water pricing) may also be used as a demand-reducing tool. Additionally, investing in technological innovation measures such as managed aquifer recharge, automatic leak detection devices, and remote sensing technology to evaluate consumption levels can help mitigate unsustainable levels of water use. Alternative agricultural practices such as switching to no-till methods that help retain soil moisture levels and investing in drought-tolerant and water-efficient crops can also substantially mitigate unsustainable groundwater use.

Key Dimension 5: Resilience to Water-Related Disasters

Data collected by the Centre for Research on the Epidemiology of Disasters show that nearly 5 billion people in 49 ADB members were affected by water-related disasters during 1990–2019, an average of 170 million people a year. Among those affected, more than 480,000 lives were lost. Compounding these massive human losses are the economic losses to assets, making it difficult for those affected by disaster to return to their lives. Another issue is the lack of research on climate change effects (Box 12). The following policy recommendations mitigate these risks and are particularly relevant in environmental water security.

Invest in disaster risk reduction infrastructure. Water infrastructure is capital intensive, with finance necessary to cover upfront construction costs and ongoing maintenance typically repaid over long periods. Investment in water-related disaster security, especially in nature-based solutions, reduces damages, benefiting both the public and private sectors. However, many of these benefits cannot be easily monetized, undermining potential revenue flows and project viability. Maintaining current levels of protection is insufficient in many countries. Investment flowing toward water-related disaster infrastructure is inadequate to bridge the gap between the investment and the requirement. Due to climate change, further investment in water-related disaster risk prevention and protection will be required to maintain current protection levels. Public finance for water-related disaster infrastructure should increase, and governments should leverage alternative financing sources by crowding in commercial finance, climate adaptation financing, and blended financing by mobilizing private sector financial resources.

Box 12: Water Security and Climate Change

Asia and the Pacific is extremely vulnerable to climate change impacts. Unabated warming could significantly undo previous achievements of economic development and improvements in living standards. At the same time, the region has both the economic capacity and weight of influence to change the present fossil fuel-based development pathway and curb global emissions. The assessment carried out by the Asian Development Bank on the regional implications of the latest projections of changes in climate conditions over Asia and the Pacific concludes that, even under the Paris consensus scenario in which global warming is limited to 1.5°C–2°C above preindustrial levels, some of the land area, ecosystems, and socioeconomic sectors will be significantly affected by climate change impacts, to which policy makers and the investment community need to adapt. However, under a business-as-usual scenario, which will cause a global mean temperature rise of over 4°C by the end of this century, the possibilities for adaptation are drastically reduced. Climate change impacts, such as the deterioration of the Asian water towers, prolonged heat waves, coastal sea level rise, and changes in rainfall patterns could disrupt ecosystem services and lead to severe effects on livelihoods, which in turn would affect human health, migration dynamics, and the potential for conflicts. The assessment also underlines that research on climate change effects is still lacking for many areas vital to the region's economy.

Source: Asian Development Bank. 2017. *A Region at Risk: The Human Dimensions of Climate Change in Asia and the Pacific.* Manila.

Address gender gaps. The tendency for water-related disasters to disproportionately affect women implies that building resilience among women across Asia and the Pacific is a key step toward disaster security in the region. Identifying gender-specific interventions, strategies, and plans of action at national and subnational governments is necessary to protect women from water-related disasters. Governments have a role to play in ensuring women have access to market-based resilience adaptations led by women themselves. Policies should aim to bolster the number of women working as scientists, economists, and policy makers and ensure that women are in leadership positions where they can enact real change.

Promote integrated flood risk mitigation, including nature-based solutions. Flood risk mapping, land use planning guidelines (restrictions on land development in flood plains), and early warning systems are cost-effective investments to limit the exposure of people and assets to water-related disaster risks. An integrated solution is important, combining structural approaches and nature-based solutions with early warning systems and residual risk instruments. Examples of nature-based solutions in this context are payment for ecosystem services to farmers in exchange for the protection of catchments and sustainable urban drainage systems.

Improve data collection and associated systems for proactive disaster risk management. A key priority in the Sendai Framework for Disaster Risk Reduction is understanding disaster risk. However, a lack of data is a constant issue in developing a clear understanding of risk. To build adequate response systems and properly budget for disaster risk, governments need to understand the scale of the issue through accurate data. Agencies need to be empowered with modern database and data collection systems so that inefficient systems predicated around reacting to disaster can be replaced with a proactive approach to disaster risk management. Satellite-based technology provides powerful tools to bridge the gap in data at a minimum cost with high efficiency.

Policy Recommendations for Water-Security Related Finance and Governance Aspects

Policy Recommendations for Governance

Addressing water-related risks in Asia and the Pacific requires robust public policies and institutions across government levels, targeting measurable objectives at the appropriate scale, relying on a clear allocation of duties among responsible authorities, and subject to regular monitoring and evaluation. Water governance can greatly contribute to designing and implementing such policies in a shared responsibility across government levels and public, private, and nonprofit stakeholders. The following policy recommendations are particularly relevant to governance and water security.

Prioritize, map, and customize stakeholder engagement. In Central and West Asia, the Advanced Economies, Southeast Asia, and East Asia, priority should be given to mapping public, private, and nonprofit actors who have a stake in water-related decisions and outcomes, including their responsibilities, core motivations, and interactions. Efforts should be made to put in place the needed formal and informal consultation mechanisms, and to customize the type and level of stakeholder engagement to the needs while keeping consultation processes flexible to adapt to changing circumstances.

Promote integrity. Central and West Asia and Southeast Asia should promote legal and institutional frameworks that hold decision-makers and stakeholders accountable, such as the right to information and investigation of water-related issues and law enforcement by independent authorities. The frameworks should include encouraging norms, codes of conduct, or charters on integrity and transparency in national or local contexts and monitoring their implementation.

Address trade-offs. In Central and West Asia, South Asia, East Asia, and the Pacific, development agendas should promote public debate on the risks and costs associated with too much, too little, or too polluted water or the lack of access to clean water and sanitation to help raise awareness, build consensus on who pays for what, and contribute to better affordability and sustainability now and in the future. This discourse should include encouraging evidence-based assessment of the distributional consequences of water-related policies on citizens, water users, and rural and urban places to guide decision-making.

Promote monitoring and evaluation. In the Advanced Economies, South Asia, and the Pacific, efforts should be made to promote dedicated institutions for monitoring and evaluation with sufficient capacity, an appropriate degree of independence, resources, and the necessary instruments to assess to what extent water policy fulfills intended outcomes. A case study in Karnataka provides examples of the many issues of monitoring systems (Box 13).

Prioritize governance financing. In the Pacific, priority should be given to promoting governance arrangements that help water institutions across government levels raise the necessary revenues to meet their mandates, including conducting sector reviews and strategic financial planning to assess short-, medium-, and long-term investment and operational needs and take measures to help ensure availability and sustainability of such finance.

Policy Recommendations for Finance

The substantial financing needs to achieve water security in Asia and the Pacific and the potential limits of prevailing financial sources (essentially domestic public finance and revenues from tariffs for water supply and sanitation services) reinforce the need to rethink typical water

Box 13: Segmentation and Asymmetries of Water-Related Information in Karnataka, India

Despite the significant efforts made by the State of Karnataka in data collection, the monitoring systems appear quite inadequate compared with the scale of the state territory and the complexity of water flow and use. Stream flows are being measured in less than 40 locations (apart from a similar number of locations monitored by the Central Water Commission). The equivalent of only one well per 200 square kilometers is being monitored. Groundwater extraction and water quality monitoring are limited, while the monitoring of aquatic ecosystems is nonexistent. Most data are not available in real time and are not publicly accessible or published. In addition, water-related data are dispersed among a wide range of sources. There are no less than seven entities responsible for collecting and producing water-related data in Karnataka, creating various issues such as data redundancy and inconsistency, data reliability and quality, and data compatibility. An integrated and harmonized database of water-related data on rainfall, geology, surface water and groundwater quantity and accuracy, water extraction and use, irrigated area, etc. would help overcome these issues, better inform policy makers, and move away from the silo approach. Streamlining state-level institutions in charge of data production into a dedicated state water resources data and information center can be considered to reduce the segmentation and asymmetries of information among water stakeholders.

Source: Organisation for Economic Co-operation and Development. 2020. *Water Governance Case Study in Karnataka.*

financing approaches. Apart from the need for huge investments, it is important to ensure that the current investments are appropriately utilized and accounted for. The following policy recommendations summarize finance policy priority toward achieving water security.

Make the best use of available assets and financial resources. Improving the operational efficiency and effectiveness of existing infrastructure and service providers can postpone investment needs by extending the operational life of existing assets. It is a prerequisite to further investment in water security by enhancing potential borrowers' creditworthiness, contributing to mobilizing additional capital.

Capturing economies of scale can contribute to this objective in rural areas or where services are fragmented and too small to secure access to technical and management skills or financial resources. Economic regulation sets performance objectives, monitors and rewards efficiency gains, and supports a robust tariff-setting process. Engagement with stakeholders sets acceptable levels of service, enhances willingness to pay, and drives water-wise behavior.

In addition, anecdotal evidence suggests that the effectiveness of expenditure programs for water security can be improved in the region. Ensuring that money is allocated to projects that deliver benefits on the ground in terms of water security can go a long way to enhancing willingness to pay and securing social support for further investments and policies that contribute to water security (Box 14). Economic analysis can help set priorities and sequence investments to maximize the benefit for communities.

Minimize future investment needs. Future investment needs can be minimized through policies that sustainably manage water resources, policy coherence, planning and setting priorities, and avoiding building future liabilities. Sequencing investments within a catchment can enable sharing the costs and benefits. Innovative technological solutions can lower costs. Properly designed, operated, and monitored decentralized systems (e.g., off-grid sanitation) and nature-based solutions can be cost-effective options with multiple benefits. Priorities in this area include (i) developing plans to enhance the water sector's long-term resilience (plans that set priorities, drive decision-making, manage uncertainties, and increase resilience);

Box 14: Reforming Dhaka Water Supply and Sewerage Authority

The Dhaka Water Supply and Sewerage Authority (DWASA) was established in 1963 to manage water supply and sewerage in the Bangladeshi capital. The WASA Act of 1996 began corporatization that ultimately professionalized DWASA and made it profitable.

DWASA had substantial water losses and poor service delivery until about 2008. Physical losses due to leakage from pipes were over 50%, and payment collection efficiency was only 62%. An Asian Development Bank (ADB) project supported a turnaround program. When the turnaround was completed in 2016, about 5.44 million people had continuous potable water supply from taps without requiring further treatment, with pressure sufficient for two-story houses.

The turnaround was anchored in infrastructure investments and policy reforms combined with visionary leadership, technical innovation, social inclusion (by supplying potable water to informal settlements), and a strong focus on public education programs and civil society involvement. In 2018, overall nonrevenue water in Dhaka had fallen to 20%, with levels of less than 10% in established district metering areas in project areas. Collection efficiency reached 97.5%, with continuous pressurized water supplied to all customers.

Sources: ADB. 2016. *Dhaka Water Supply Network Improvement Project*. Manila; DWASA; and ADB. 2020. *Asia's Journey to Prosperity: Policy, Market, and Technology over 50 Years*. Manila.

(ii) supporting plans with realistic financing strategies; (iii) encouraging policy coherence across water and other policy domains that affect water availability and demand, or exposure and vulnerability to water-related risks; (iv) managing water demand and strengthening water resource allocation; (v) developing flood risk mitigation strategies; and (vi) exploiting innovation in line with adaptive capacities.

Harness additional sources of finance. Asia's investment deficit in water supply and sanitation, flood protection, and irrigation infrastructure will require leveraging financial resources from diverse potential sources. National and local governments need to (i) increase contributions from polluters, users, and beneficiaries; (ii) increase reliance on domestic funds; (iii) seek funding from external agencies; and (iv) attract private investment. Transitioning from concessional finance to crowding in commercial capital will be crucial. Central governments have a distinctive role in setting the enabling conditions and encouraging the development of a dynamic finance industry.

Therefore, priorities for establishing new financing sources include (i) ensuring that tariffs for water services reflect the costs of service provision; (ii) considering new financial sources from users, polluters, and beneficiaries; and (iii) leveraging public finance and risk mitigation instruments (such as guarantees) to crowd in commercial finance.

Strategic Directions for ADB Water 2030: A Water-Secure and Resilient Asia and the Pacific

ADB's Strategy 2030 provides a framework to address the regional developmental challenges over the next decade. It focuses on targeting interventions to the poor and disadvantaged, supporting economic growth, integrating solutions to improve the quality of life in cities and rural areas, and achieving food security. Realizing these goals requires improving governance arrangements to deliver services, building resilience of communities and systems, achieving gender balance in access to resources, safeguarding the natural environment, and enhancing cooperation beyond borders. Integral to all operational priorities is improving water security equitably. Hence, Strategy 2030 goals and the thematic focus of its operational priorities require investments in water security and related capacity building and policy within a more integrated approach with other sectors. The financing gap is significant. Investment needs in water and sanitation alone are estimated to be on average $53 billion per year up to 2030, of which about one-third will be needed from the private sector to help fill the gap.

ADB's Water Sector Framework 2021–2030[52] sets out ADB's approach to addressing the region's water challenges to achieve Strategy 2030's overall goals. It thus links to the delivery of ADB's corporate results framework and contributes to initiatives by ADB members to achieve the SDGs and other international commitments such as the Paris Climate Agreement and associated Nationally Determined Contributions, the Convention on Biological Diversity, and the Sendai Framework for Disaster Risk Reduction.

In addition, ADB Water Sector Framework 2020–2030 aims to advance a water-secure and resilient Asia and the Pacific. It builds on the impressive advances already made, promotes an exchange of lessons and experiences, and encourages reform initiatives that incentivize sustainable outcomes. AWDO, with its multiple KDs, is a clear example of a long-term integrative and multisector knowledge, policy, and communication partnership between ADB and its developing members. AWDO has been used to inform the framework and constitutes a key delivery assessment mechanism toward the Strategic Directions for Water[52] and ADB's Strategy 2030.

ADB's Water Sector Framework 2020–2030

The Water Sector Framework 2020–2030 will see ADB intensify its support for a water-secure and resilient Asia and the Pacific—the framework's vision—through inclusive, well-governed, and sustainable water services and resource management in accordance with Strategy 2030. The framework comprises five guiding principles to inform ADB's programming and project planning in each of the four focal areas for delivering successful and sustainable outcomes (Figure 31). Implementing the framework will lead to better served, healthier, and more prosperous communities.

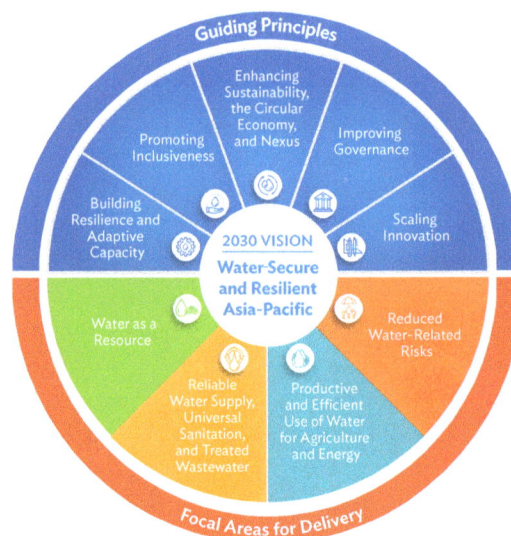

Figure 31: Guiding Principles and Focal Areas for Delivery – Strategic Directions for ADB Water 2030

Source: Asian Development Bank.

---

52 ADB. Forthcoming. Water Sector Framework 2021-2030: Water-Secure and Resilient Asia Pacific (working title).

Guiding Principles for Improving Development Outcomes

Building resilience and adaptive capacity. Addressing risk and uncertainty of climate change, water-related disasters, public health emergencies, and other economic shocks is central to the next 10 years of ADB's water program. Water is inherent to the human, environmental, and economic dimensions of climate adaptation. Increasing occurrences of extreme flood, drought, and tropical cyclones require an intensified focus on resilience and climate proofing in projects, including water infrastructure resilience. Strengthening community resilience to water security risks and the ability to "build back better" is necessary to mitigate climate impacts and other financial, environmental, and health shocks, including resource scarcity, system capacity constraints, and poor water quality.

Promoting inclusiveness. Water projects have the potential to be models of inclusiveness for access to services, use of resources, gender equality, and livelihood opportunities. ADB will promote engagement of groups that have been marginalized from project benefits and excluded from involvement in the planning and design process. Women have long felt the burden of poor water service coverage, and initiatives and continued efforts will target improvements through water, sanitation, and hygiene programs. Access to resources and support mechanisms will be needed to increase the number of households headed by women.

Embracing sustainability. Integrated approaches offer new opportunities to achieve global sustainability goals, including a greater focus on ecosystem services, the circular economy, and linkages between sectors. These approaches require initiatives that break down conventional barriers, improving resource efficiency and treating waste as a resource by incentivizing recycling and reuse. Adopting nature-based solutions and focusing on green infrastructure will help restore ecosystems, producing multiple benefits in rural and urban

settings. The nexus between water, food, and energy security involves public policy and project design that recognize interdependencies and trade-offs, encourage environmental sustainability, and reduce carbon emissions.

Improving governance. Good governance is critical for delivering effective services across all water subsectors and the sustainable management of water. ADB will support a long-term programmatic approach, rigorous diagnostics through sector analysis to assess and improve institutional capacity and resources in counterpart agencies, and measures to address political and socioeconomic risks. Strategy 2030 highlights a range of governance interventions, including policy and regulatory frameworks, improved public investment management, and sustainable finance arrangements such as tariff policies, appropriate business models, improved project management skills, and sustainable asset operation and maintenance.

Fostering innovation. The rapidly evolving digital and technological advances are redefining the possibility for water resources management and service delivery. ADB will encourage and scale up innovative technologies and digital solutions across the water sector, including smart network management, remote sensing and geographic information system, real-time data generation, and digital payment system (Box 15). Greater access to accurate information also introduces opportunities for new levels of oversight by water users. Similarly, ADB will encourage innovative financing arrangements that offer new opportunities for project implementation.

Focal Areas for Delivering ADB Support

Four focal areas represent ADB's water operations interface with its developing member countries (DMCs). Through these entry points and the five guiding principles, ADB actively supports integration and coherence across subsectors to deliver sustainable outcomes.

Box 15: Smart Management of Rural Drinking Water Services in West Bengal

According to the World Health Organization (WHO), about 140,000 Indians died in 2016 from diarrhea due to inadequate water. The Government of India is implementing new approaches and has made a concerted effort by launching the ambitious Jal Jeevan Mission in 2019 to provide water for every tap by 2024 to rural households to prevent avoidable deaths. A shift from business as usual is required from all sectors, including citizens, to achieve the Sustainable Development Goals (SDGs) and improve rural water service delivery.

The Asian Development Bank (ADB) supports the Government of West Bengal, through the West Bengal Drinking Water Sector Improvement Project (WBDWSIP), in three districts suffering from groundwater contamination with fluoride, salinity, and arsenic: Bankura, East Medinipur, and North 24 Parganas. The population is further exposed to natural disasters like riverine flooding, storm surges, and climate change impacts. Through the WBDWSIP, around 2 million people will receive 24/7 piped water at their households from 2022.

To improve system efficiency, strengthen customer service, build resilience, and secure public participation, WBDWSIP introduces many innovative approaches and digital technology-driven smart water management. The source of water upstream of treatment plants will be monitored through the Internet of Things (IoT) technology with inexpensive sensors readily connected to a central database, where data are further analyzed and applied for low flow (or flood) and salinity forecast and early warning. This approach will help West Bengal's Public Health Engineering Department (PHED) manage water intakes to ensure overall water quality and security, e.g., from the estuaries with shifting freshwater and saline water using pumps and intermediate storages.

Leakage detection and water quality monitoring in the network are handled by a combination of real-time sensors, the IoT, and reporting via mobile phones among operators and consumers using the Internet of People (IoP). Grievance and fault reporting apps on smartphones for improved services and asset management are being developed and tested in villages. This development lends well to WBDWSIP's innovative and inclusive service delivery model. The bulk facilities will be managed by PHED and distribution network and services within the villages by the respective gram panchayats, over 30% of whom will be women.

Using the IoT and IoP and a participatory approach with villagers and service providers represents a huge potential for water management in rural India, where mobile phones can serve as the main interface connecting people to the internet, whether it is end consumers in rural areas or operators in the large water supply schemes. This approach is a significant innovation for rural water service delivery in developing countries, the first in India for large-scale rural water schemes.

Sources: World Health Organization. Global Health Observatory Data Repository. Burden of Disease (accessed 26 October 2020); Government of West Bengal, Public Health Engineering Department; and Asian Development Bank.

Water as a sustainable resource. ADB's policy initiatives will continue to promote regulatory and incentive mechanisms for improved water sector governance and sustainable management of surface water and groundwater resources. Addressing linkages from the water source to coastal areas will result in economic growth opportunities, restored aquatic ecosystems, and improved livelihoods. Good practices of sustainable resource management and resilience from emerging economies in the region can be models for others through South–South cooperation, including reversing the water quality impacts of rapid economic growth.[53] Two or more states share many

---

[53] South–South cooperation refers to technical cooperation (exchange of resources, knowledge, skills, and technology) among developing countries in the Global South. United Nations. 2019. *What Is 'South-South Cooperation' and Why Does It Matter?* News. 20 March.

of the region's rivers and aquifers. ADB supports regional cooperation platforms as an opportunity for engagement on a range of water-related challenges.

Universal and safe water services. Reliable water supply and universal sanitation for urban and rural users and sustainable wastewater management remain major areas for ADB support to address unmet demand, particularly among marginalized communities. With the current COVID-19 pandemic, water, sanitation, and hygiene (WASH) services prove to be an effective response. ADB supports access to adequate and inclusive sanitation systems appropriate to local conditions, including the Citywide Inclusive Sanitation flagship program, through sewered and non-sewered, and centralized and decentralized systems. Infrastructure investments will increasingly be planned within broader spatial and urban planning objectives, promoting circular economy principles by recognizing waste as a resource, providing a greater emphasis on achieving health security and incorporating resilience to climate risks. ADB will support utilities to improve water services performance through reduced cost, sustainable finance, reduced nonrevenue water, and enhanced asset management. ADB will continue to assist its DMCs in providing an enabling environment for private sector involvement through policies and regulatory frameworks that clarify respective roles and risk apportionment.

Productive water in agriculture and energy. The contribution of irrigated agriculture to food security and rural livelihoods is significant for ADB's agriculture portfolio. To accelerate rural revitalization and climate adaptation, ADB's irrigation investments will support diversified and higher value agriculture, ensure more efficient and productive water use, target benefits to the rural poor and marginalized, and promote the shift to low carbon and diverse food systems by adopting modern irrigation practices and institutional reforms. This approach also requires projects that are better integrated into a value chain approach. ADB operations will increasingly seek to ensure compatibility between energy and water resource planning and will further explore support to promote renewable energy and higher efficiencies in water and energy projects.

Reduced water-related risks. ADB will continue to build on its investments to reduce disaster risk and enhance resilience. Disaster risk reduction interventions will be integrated with other development programs, including livable cities and food security, by encouraging a mix of structural and nonstructural measures such as infrastructure, nature-based solutions, and risk-sensitive land use management approaches. Recognizing the need for a more holistic perspective, ADB supports integrated flood risk management approaches to shift the emphasis from disaster management to disaster risk management. Clear strategies will increasingly be needed to deal with recurrent droughts.

Sewer network rehabilitation and expansion works at the Malakal and Meyuns area in Palau

PART IV
AWDO and Policy in Action

Karnataka, India: A Subnational-Level Application of AWDO

Timor-Leste: Applying AWDO to Determine Investment Needs

Thailand: A National-Level Application to Support the National Strategic Master Plan

Yellow River Water Sector Assessment and AWDO: Innovation and Future Application

Lessons from Earlier AWDO Country Applications: Bhutan, Mongolia, and the People's Republic of China

The Asian Water Development Outlook (AWDO) methodology provides an overview of water security and policy formulation in Asia and the Pacific, enabling a comparison over time between ADB members and regions. Based on the assessment in Block II, including the Organisation for Economic Co-operation and Development (OECD) studies, conclusions and recommendations are provided in Block III.

In addition to its application at the level of ADB members, the AWDO methodology can also be used at a more granular level in a geographic sense, as highlighted below by some subregional case studies where AWDO is applied to improve water security and the associated water resources policy.

In the AWDO 2020 framework, three country assessments have been carried out: Thailand, Karnataka (India), and Timor-Leste. Apart from these three case studies, a planned future application at the basin level (i.e., the PRC's Yellow River basin) is also discussed in this section as well as the assessment of earlier applications of the AWDO methodology in Bhutan, Mongolia, and the PRC.

Karnataka, India: A Subnational-Level Application of AWDO

Karnataka's water demand is expected to double by 2030. A business-as-usual approach will make it difficult to meet increasing water demand, especially since Karnataka is one of India's most water-stressed states. About 58% of Karnataka's territory is drought prone, and the state experienced severe drought conditions yearly during 2011–2016. Water scarcity and associated competing water uses (among agriculture, industry, and households) and other economic uses can be observed in the tensions and disputes settled by tribunals despite clear priorities for water allocation set by the State of Karnataka.

The Government of Karnataka is responsible for managing these water resources at large. To assist the government, the Advanced Centre for Integrated Water Resources Management—a think tank to the government's Water Resources Department, which was established with ADB support—has applied AWDO 2016 methodology at the state level. Recognizing integrated water resources management (IWRM) as an adaptive process, the think tank leads institutional change and capacity development processes while building a sector knowledge base. This state initiative is happening in the context of a countrywide renewed thrust on water-related aspects ranging from drinking water to improved sanitation to groundwater recharge.

The five AWDO 2016 water security indicators will be used as benchmarks in the upcoming state water policy, currently in its final stages of approval. Future trajectories can thus be charted based on these indicators and associated goals to guide sustainable water resources management while ensuring the much-required water security in Karnataka. The application of AWDO at the district and/or city level will also help water security assessment and program planning and design accordingly. Periodic assessment of AWDO, including the future application of the AWDO 2020 methodology, measures, monitors, and reviews the progress of government schemes and programs to achieve better water security for a sustainable future.

Following the OECD Principles on Water Governance, the government assessed water governance in Karnataka and shed light on four governance challenges:

(i) **Fragmentation of roles and responsibilities.** Although the Water Resources Department acts as the main state institution for water policy, there is a plethora of other stakeholders involved in and/or responsible for irrigation projects, groundwater management, pollution control, drainage, water and sanitation works, etc. This crowded institutional setting—where water management has been structured into silos for groundwater, surface water, irrigation, and domestic use, with little dialogue across silos—creates conflicting objectives, leading to a condition called "hydroschizophrenia."[54]

(ii) **Scattered and patchy data and information.** Despite the significant efforts made by the State of Karnataka in data collection, the monitoring systems appear quite inadequate compared with the scale of the state territory and the complexity of water flow and use. Groundwater extraction and water quality monitoring are limited, while the monitoring of aquatic ecosystems is nonexistent. Most data are not available in real time, not publicly accessible or published, and dispersed among a wide range of sources.

(iii) **Scale mismatch for integrated water resources management.** Although Karnataka is covering seven river basins, only two IWRM structures have been created in the state.

(iv) **Low level of stakeholder engagement.** Stakeholder engagement is institutionally prescribed for some water management aspects, such as irrigation with the Irrigation Consultative Committees. However, in practice, water users are poorly engaged in planning, managing, and controlling water resources.

The governance case study analysis in Karnataka highlighted the following policy recommendations to address identified governance challenges:

(i) Bearing in mind the major challenges related to groundwater depletion and decreasing rainfall, overcoming the fragmentation of responsibilities for irrigation (which represents 85% of total water use) is essential to address sustainable water resources management. Streamlining functions and mandates among existing water resources management institutions could be a way forward.

(ii) An integrated and harmonized database would help better inform policy makers and move away from the silo approach. Streamlining state-level institutions in charge of data production into a dedicated state water resources data and information center could be considered to reduce the segmentation and asymmetries of information among water stakeholders and ensure a more comprehensive and systematic approach to data collection and analysis.

(iii) Implementing river basin governance arrangements to all river basins within Karnataka could ensure coordination between stakeholders and across riparian states for effective water resources management at the appropriate scale. This approach could transform water use conflict resolution into effective and integrated basin planning, and water crisis management into water risk management.

(iv) Further stakeholder engagement in water decision-making should be developed through various approaches depending on the intention pursued.

It is noted that the conclusions and recommendations given above are based on the specific case study on Karnataka and should not be regarded as general conclusions for India. At national level, many initiatives are ongoing that address some of the issues raised in the Karnataka case study.

Timor-Leste: Applying AWDO to Determine Investment Needs

With an overwhelmingly young population of 1.3 million, which is rapidly growing at 2.4% annually, Timor-Leste is the newest nation-state in Southeast Asia. It has made gains to reduce poverty from 47% in 2007 to 23.1% in 2018.[55] Since its independence

54   Jarvis, T. et al. 2005. International Borders, Ground Water Flow, and Hydroschizophrenia. *Ground Water*. 43 (5). pp. 764–770.
55   World Bank. March 2020. Data Catalog. https://datacatalog.worldbank.org/dataset/world-development-indicators.

in 1999, after decades of conflict, it has improved the lives of many people.

Timor-Leste has ample water resources with 8.125×10^9 cubic meters per year[56] for human, economic, and environmental development. However, water availability is highly seasonal due to Timor-Leste's tropical monsoon climate, making water availability unreliable, geographically uneven, or poorly understood (as is the case for groundwater). Timor-Leste experiences both regular droughts and floods, negatively impacting the majority of the rural population (68%), who rely on subsistence agriculture. Timor-Leste's water infrastructure mainly consists of piped water networks in the capital and municipal centers and irrigation networks for rice production, much of which still require rehabilitation. While access to water supplies is estimated at 98% for urban and 70% for rural populations, these figures do not consider the quality, reliability, and sustainability of services.[57] Access to sanitation is a policy focus for the government presently being only 76% for urban and 44% for rural populations.[57]

As part of an ADB-funded technical assistance project to Timor-Leste to strengthen its water sector, the AWDO comprehensive framework is currently being applied to assess water security based on the best available information across its five KDs (rural, economic, urban, environmental, and disaster securities) in Timor-Leste at a subnational level (i.e., in Timor-Leste's 12 municipalities and the special administrative region of Oé-Cusse Ambeno). While national data are mostly available, a paucity of data makes subnational level analysis challenging, with only 47% of the data necessary to calculate all KDs are available at municipal level.

The Government of Timor-Leste approved both the National Public Water Supply Policy and the National Water Resources Management Policy in 2020. The policies outline national responsibilities,

intentions, objectives, and strategies for managing water supply and water resources and providing 2030 frameworks for action. When the AWDO framework is finalized based on available data, it will give an overview of rural and urban water security (KD1 and KD3) and analyses of KD2, KD4, and KD5. The initial assessment already shows that some municipalities, such as Oé-Cusse Ambeno, lack water security, especially for rural household water and sanitation services. The analysis will be useful to highlight inequalities and vulnerabilities between municipalities and make recommendations accordingly. The findings and recommendations will assist action planning to implement the newly approved policies, which will be used to engage with relevant ministries to support collaborative, evidence-based dialogue and help align decision-making and planning for water resources and infrastructure with Timor-Leste's Strategic Development Plan, 2011–2030. By presenting water security as an issue of concern to multiple ministries, AWDO will open a platform to pursue interministerial dialogue and build a consensus on addressing water security issues in Timor-Leste.

The AWDO assessment will help prioritize actions and make recommendations on filling key data, policy, and investment gaps, keeping available resources and constraints in mind. The assessment may also prioritize water resource assessments to understand sustainable yields and develop river basin plans.

Following the OECD Principles on Water Governance, the government assessed water governance in Timor-Leste and shed light on three governance challenges that hinder the effectiveness of water policies in the country:

(i)   Although the Decree Law No. 6/2015 allocates water resources management and water supply and sanitation responsibilities among various ministries, some overlaps between institutions can be observed and

---

[56]   FAO. AQUASTAT - FAO's Global Information System on Water and Agriculture. http://www.fao.org/aquastat/en/databases/maindatabase/ (accessed 1 November 2020).

[57]   JMP. 2017. WHO/UNICEF Joint Monitoring Programme for Water Supply, Sanitation and Hygiene (JMP). https://washdata.org/.

can generate competition and conflicting policy objectives. With the National Public Water Supply Policy and National Water Resources Policy approved, the country has a clearer policy vision for water resources management and water supply and sanitation. Much work remains to be done to communicate and share this vision with all stakeholders, set out clear improvement targets, develop an action plan, and monitor achievement.

(ii) The funding gap appears as one of the main obstacles to effective implementation of Timor-Leste's Strategic Development Plan and SDG 6 despite considerable needs in the context of population and urbanization growth.

(iii) Government capacity is a key constraint on the implementation and success of water sector initiatives, with a lack of staff with water-related technical knowledge, expertise, and managerial competences hampering water policies implementation.

The governance case study analysis in Timor-Leste highlighted the following policy recommendations to address identified governance challenges:

(i) A concrete action plan clarifying operational steps (time frame, resources required, milestones, responsible institutions for delivering the action, etc.) is necessary to reach policy goals. The action plan should also be reviewed periodically to ensure that policy implementation and improvements on the ground are effectively happening as planned.

(ii) As identified in the National Water Resources Policy, a funding scheme based on the collection of water abstraction and/ or effluent discharge fees could be put in place at the national or district level to ensure the long-term funding of water resources management policies.

(iii) A possible way to start addressing the capacity gap is to include dedicated capacity building and development in sector strategies and policies. Improving

public procurement and tendering capacities, as well as competencies to design and manage investment projects, would also be required to ensure better implementation of water-related policies.

Thailand: A National-Level Application to Support the National Strategic Master Plan

Water security is of utmost importance to Thailand due to competing and increasing water demands in the agricultural, industrial, and service sectors; deteriorating natural water resources caused by pollution; greater damages from floods and droughts due to climate change; and the challenges presented by the water–energy–food nexus, transboundary river and aquifer management, and fragmented institutional framework for water resources management (Box 16).

Thailand's National Strategy, 2018–2037 provides a framework toward security, prosperity, and sustainability for all and mainstreams water security as part of its eco-friendly development and growth strategy. Specifically, Thailand aims to create eco-friendly water, energy, and agricultural security by (i) developing the entire river basin management system, (ii) enhancing water system productivity through efficiency and value-added generation, (iii) creating a national energy strategy, (iv) enhancing energy efficiency, and (v) developing agricultural and food security at national and community levels.

The national strategy is implemented through a series of master plans at different levels, which include the National Strategic Master Plan and the Water Resources Master Plan developed by the Office of the National Water Resources. Under the National Strategic Master Plan, the national strategy drafting committees have adopted the SDGs and the AWDO framework (KD1–KD5) as well as water productivity and water governance as key indicators for their water security policy.

Box 16: Water Management in Thailand in the Face of Droughts and Floods

Water management in Thailand is characterized by a highly fragmented institutional framework consisting of at least 31 ministerial departments under 10 ministries, 1 national unit under the Prime Minister's Office, 1 agency, and 6 national committees. Overlapping responsibilities can lead to conflicts of interest and impede the development of integrated water management. Unlike many countries, Thailand has no single law governing water management. Currently, there are 36 primary laws and 2,000 secondary legal frameworks relating to water resources management. For this reason, the Department of Water Resources has been working since 1992 to draft the Water Act. This act aims to rationalize the legal framework, strengthen existing legal instruments, and ensure the effectiveness of policies. In 2017 the Office of the National Water Resources (ONWR) was formed. The new Water Resources Act has come into force in 2018. By providing policy guidance and setting homogeneous national priorities, it is intended to allow different entities and stakeholders to develop and implement their respective water management plans in accordance with the overarching national framework. Budget allocation will also be compliant with agreed national priorities. Good practice of unified management will be drawn from the Ministry of Energy's experience.

Without a comprehensive law, Thailand has launched some water management plans and strategies. For example, the Water Resources Management Strategy, 2015–2026 covers water source management, water usage, and wastewater management. The Strategy for Green Growth under the 20-Year National Strategy Framework, 2017–2036 and the Strategy for Green Growth toward Sustainable Development under the 12th National Economic and Social Development Plan, 2017–2021 foresee numerous activities related to water management. However, implementation is hampered by institutional complexity and political issues. Furthermore, Thailand's water management plans cover a relatively short period. Longer-term projections and planning are needed, incorporating factors that influence the probability of future floods, such as rising sea levels and land subsidence.

After the floods in the northeast region in August 2017, the Prime Minister established the National Water Management Unit under his office to oversee government efforts to tackle flooding and droughts across relevant ministries and government agencies. The unit's operational success, its relationship with existing bodies, and its effectiveness in addressing the challenges outlined above remain to be seen.

The response to disasters such as flooding also falls under Thailand's disaster risk governance frameworks. Taking the Sendai Framework for Disaster Risk Reduction, 2015–2030 as a set of guiding principles, the government adopted the National Disaster Risk Management Plan in 2015, supplementing the Disaster Prevention and Mitigation Act 2007. One of the plan's goals is to improve coordination—an identified weakness of previous approaches—between the different parts and levels of government responsible for disaster management. The plan refers to nonstructural mitigation measures such as land use planning, zoning, building codes, and other incentive measures, also lacking in previous approaches. These measures transform a reactive disaster response and recovery mode into a proactive approach that encompasses disaster risk reduction. Combining disaster risk management with climate change adaptation plans would further increase Thailand's resilience to disasters.

Sufficient funding and increased capacity at the local level will be needed if plans are to be effective. Currently, a lack of oversight means that funds transferred to the local level can be diverted for other purposes. Effective response to disasters is also hampered by the incomplete decentralization of disaster governance. Local authority organizations (except Bangkok Metropolitan Area) lack the capacity to respond effectively to disasters and receive insufficient assistance from the central government.

Source: Organisation for Economic Co-operation and Development (OECD). 2018. *Multi-Dimensional Review of Thailand (Volume 1): Initial Assessment*. Paris: OECD Development Pathways, OECD Publishing.

In addition, Chulalongkorn University applied a modified AWDO 2016 framework to assess Thailand's water security status at provincial and river basin scales. This preliminary study highlights that urban and environmental water security still presents challenges and that resilience to water-related disasters remains low in Thailand. The study results could be further developed into a multilevel and multisector platform permitting alignment of policies, budget, planning, and implementation of interdisciplinary approaches toward water security.

Yellow River Water Sector Assessment and AWDO: Innovation and Future Application

Considered the cradle of Chinese civilization, the Yellow River gets its name from the loess (fine-grained yellow sediments) suspended in the water. Running 5,464 kilometers from its source in Bayan Har Mountains on the Tibetan Plateau to the Bohai Sea, it is the second-largest river in the People's Republic of China (PRC), after the Yangtze River. It covers a basin area of 752,400 square kilometers, crosses nine provinces, and is home to around 120 million people. Underpinning the basin's economic and social development, the river is crucial to sustaining people's livelihoods and supplies water to 66 prefectural level cites and 340 counties. The basin is the PRC's "food basket," accounting for 8% of national GDP. Agriculture output contributes to feeding about 12% of the population and irrigates about 15% of arable land. The river also faces several water security challenges.

The first challenge is related to flood risks. Millions of lives have been lost due to flood disasters over the centuries. From 206 BC to AD 1949, 1,092 major floods were recorded, along with 1,500 dike failures, and 26 river rechanneling projects. Since 1949, master planning for flood control and construction of numerous hydraulic structures has significantly reduced the vulnerability and losses due to floods. To date, approximately 90 million inhabitants live in the

flood-prone area in the flat North China Plain, while 1.9 million people living in the inner flood plain of the lower Yellow River are still facing imminent threat.

The second challenge is water scarcity. The basin feeds about 12% of the country's population with only 2% of the total water resources. It experienced several droughts, culminating in 1997 with the largest number of zero flow days (226 days). Physical water shortage is exacerbated by climate change, with the total water resources shrinking by 13.4% since 1990. Natural resources are unevenly distributed along the basin, with water resources concentrated in the upper reaches of Lanzhou, accounting for about 60% of the basin's water. The cultivated land is mainly in Hetao, Fenwei, and lower plain areas, which take up 67% of the total cultivated land areas and are characterized by better irrigation conditions. Energy and mineral resources are mostly located in the middle and lower reaches, where coal reserves account for more than 50% of the country. Balancing scarce water resources among competing uses is difficult, with the agriculture sector withdrawing about 69.6% of water.

The third issue is sediment load. The Yellow River is one of the most hydrologically complex systems in the world, although the amount of sediment flows has decreased in recent years. Under natural conditions during 1919–1959, the amount of sediment in the Yellow River was 1.6 billion tons. This generated the phenomenon of the "hanging river," with the bed reaching 10 meters above the adjacent plain in the downstream valley. During 2000–2018, the average annual sediment transport volume at Tongguan Hydrological Station decreased to 277 million tons. The sediment is mainly from the middle reaches of the Loess Plateau, with the midstream sediment transport accounting for 93% and the surface water resources being 36%. This issue is compounded by soil erosion and flood risks, the biggest threat in the basin.

Last, water resources are further strained due to pollution from unsustainable agriculture practices, mining, local industries, and untreated wastewater from rural and urban areas. Three-quarters of the total basin area is considered ecologically fragile, with 75% of grasslands degraded and 20% of groundwater resources overexploited.

Against this backdrop, ADB conducts a water sector assessment of the Yellow River basin, adapting the Asian Water Development Outlook (AWDO) 2020 methodology across its five KDs of water security. The assessment is aimed at informing ADB's forthcoming Yellow River ecological corridor program to support the PRC's goal toward achieving high-quality development. To carry out the water sector assessment, ADB will be closely working with the Yellow River Conservancy Commission of the Ministry of Water Resources and the Ministry of Ecology and Environment. The study will also engage provincial and municipal governments in the upper and middle reaches, which are highly affected by water and ecological security issues. Based on the results and the analysis conducted, policy recommendations will be drafted and submitted to the Ministry of Finance and the National Development and Reform Commission to help inform and prepare ADB's lending pipeline for project investments in the Yellow River basin. In conducting the assessment, ADB will apply the know-how adopted in the Yangtze River Economic Belt Program and build on the previously conducted water sector assessment of the PRC, where AWDO methodology was used and adapted at a national scale.

Lessons from Earlier AWDO Country Applications: Bhutan, Mongolia, and the People's Republic of China

The AWDO methodology is developed for all ADB members, which are different in water resource system conditions, size and population (e.g., the PRC versus a Pacific island state), and climatological conditions (e.g., Mongolia versus Indonesia). Differences might

also come from political foci of members on KDs that are hardly or not at all relevant to a certain member country (e.g., KD1 in Singapore). The different conditions generate country-specific adjustments to the methodology. Examples of such tailor-made applications are the following.

Bhutan applied the AWDO methodology to produce the Bhutan Water Security Index. While the AWDO KDs were maintained, the underlying indicators were tailored to Bhutan's conditions and readily available and government-collected information. The Bhutan Water Security Index has been formulated not just as a top-down monitoring tool but also as a basis for planning with dimensions and indicators directly used in a logical framework analysis and river basin planning.[58] Water security was assessed for the whole country and at the watershed level. The AWDO application has contributed to the development of Bhutan's IWRM.

The Mongolia Country Water Security Assessment[59] is based on the AWDO 2016 methodology. With the extensive stakeholder consultations in Mongolia, the AWDO 2016 methodology was customized to the Mongolian context and piloted in five river basins. The assessment was conducted for all 29 river basins in the country, incorporating lessons learned from the previous assessments. The adaptation included the redefinition of KD1 to rural households (now also applied in AWDO 2020); the special attention given in KD2 to the country's specific livestock conditions; and the inclusion of *dzud* (long, strong winter conditions) in KD5, replacing the sub-indicator storm surges. With an in-depth institutional analysis at the river basin scale, the AWDO application in Mongolia was used to prepare a water sector investment program.

The AWDO methodology was applied at the PRC's 31 provinces and 10 river basins.[60] It was adjusted to align the approach more to political priorities. KD3

---

[58]   ADB and Government of Bhutan, National Environment Commission. 2016. *Water: Securing Bhutan's Future*. Thimphu.

[59]   ADB. 2020. *Overview of Mongolia's Water Resources System and Management: A Country Water Security Assessment*. Manila.

[60]   ADB. 2018. *Managing Water Resources for Sustainable Socioeconomic Development: A Country Water Assessment for the People's Republic of China*. Manila.

(urban water security) was combined with KD1 (rural household water security). In addition to KD4 (environmental water security), a KD on ecological water security was introduced, stressing the importance of the water system's sustainable ecological health and sustainable ecological services.

In conclusion, the AWDO methodology can be applied at the subnational (province, river basin) level in the water security assessments, providing supporting information to develop water resources policies and plans and investment needs. The methodology is sometimes adjusted to local conditions to make the results more policy relevant at national and subnational levels, without losing the strength of the overall approach.

Aerial view of Phong Nha-Kẻ Bàng National Park in Quang Binh province in Viet Nam

PART V
Appendixes

APPENDIX 1
National Water Security Index

Scoring Approach of Key Dimensions for National Water Security

For each key dimension (KD), a specific scoring approach has been developed based on the indicators used for that KD, resulting in different score tables. The following are the maximum scores of each KD: 20 for KD1 (rural household water security), 20 for KD2 (economic water security), 17 for KD3 (urban water security), 10 for KD4 (environmental water security), and 15 for KD5 (water-related disaster security). To make these scores comparable, the scores of each KD have been "normalized" to a maximum of 20 by multiplying the actual score by a factor 20/

(maximum score). The tables in these appendixes list both actual and normalized scores. The KD figures presented in this report are all based on the max-20 scores. The national water security (NWS) score is the sum of the max-20 scores of the five KDs—a maximum score of 100.

The NWS index (development stage) on a scale of 1–5 is derived from the scores of Table 2 of the main report. An index of 1 (NWS score < 42) expresses that water security in that specific country is "nascent," while an index of 5 (NWS score ≥ 96) means that the country is a "model" that has achieved water security. Table A1 presents the KD and NWS scores for all 49 ADB members.

Table A1: Scores and National Water Security Index

Economy	KD1	KD2	KD3	KD4	KD5	NWSI Score	NWSI
Scale	1–20	1–20	1–20	1–20	1–20	1–100	1–5
Afghanistan	4.0	7.5	10.0	10.6	7.3	39.5	1
Armenia	17.0	11.6	14.7	11.9	16.6	71.8	3
Australia	19.0	15.7	17.6	15.4	20.0	87.8	4
Azerbaijan	11.0	14.4	14.7	10.1	14.9	65.1	3
Bangladesh	9.0	11.8	12.1	9.0	11.0	52.8	2
Bhutan	10.0	11.2	13.8	13.9	13.9	62.8	3
Brunei Darussalam	14.0	14.3	12.6	15.5	19.1	75.5	3
Cambodia	9.0	10.0	13.2	15.4	9.8	57.5	2
China, People's Republic of	14.0	16.4	17.9	10.8	13.5	72.7	3
Cook Islands	18.0	9.0	10.6	18.8	16.0	72.5	3
Fiji	12.0	13.6	9.1	12.5	12.3	59.5	2
Georgia	14.0	10.0	13.2	9.5	17.7	64.4	3
Hong Kong, China	N.A.	15.3	19.4	12.0	18.8	81.5*	4
India	6.0	10.9	9.7	8.1	12.1	46.8	2
Indonesia	11.0	12.2	11.5	14.4	11.9	61.0	3

continued on next page

Table A1 continued

Economy	KD1	KD2	KD3	KD4	KD5	NWSI Score	NWSI
Scale	1–20	1–20	1–20	1–20	1–20	1–100	1–5
Japan	19.0	14.3	18.5	16.5	19.6	87.9	4
Kazakhstan	16.0	16.1	15.9	9.4	16.3	73.7	3
Kiribati	7.0	11.3	5.9	16.2	7.9	48.2	2
Korea, Republic of	20.0	15.6	18.2	10.4	19.7	84.0	4
Kyrgyz Republic	13.0	10.8	16.2	15.0	17.6	72.6	3
Lao People's Democratic Republic	7.0	9.2	15.9	13.0	10.1	55.2	2
Malaysia	14.0	15.5	11.5	15.4	18.2	74.7	3
Maldives	20.0	10.7	9.7	7.6	14.8	62.8	3
Marshall Islands	9.0	7.0	9.1	15.1	8.6	48.9	2
Micronesia, Federated States of	11.0	6.0	4.7	13.0	7.4	42.0	2
Mongolia	8.0	10.9	11.5	13.2	17.5	61.1	3
Myanmar	6.0	10.8	9.7	11.9	10.2	48.6	2
Nauru	N.A.	7.0	7.1	17.6	15.7	58.6*	2
Nepal	6.0	9.7	11.2	13.1	12.4	52.3	2
New Zealand	20.0	14.9	19.4	14.7	20.0	89.1	4
Niue	15.0	4.0	10.3	13.3	18.4	61.0	3
Pakistan	6.0	9.9	10.3	7.7	8.8	42.7	2
Palau	19.0	9.7	15.3	12.1	16.9	73.0	3
Papua New Guinea	4.0	8.8	5.6	12.5	12.0	42.8	2
Philippines	11.0	11.3	16.8	16.7	12.1	67.8	3
Samoa	14.0	10.3	12.6	13.8	12.0	62.8	3
Singapore	N.A.	15.4	19.1	7.8	20.0	78.0*	4
Solomon Islands	7.0	14.0	7.9	9.5	10.9	49.3	2
Sri Lanka	14.0	10.0	14.4	9.7	11.8	60.0	2
Taipei,China	17.7	14.2	16.5	14.1	18.3	80.8	4
Tajikistan	9.0	8.4	11.2	12.1	17.4	58.1	2
Thailand	14.0	14.7	9.1	10.1	10.8	58.6	2
Timor-Leste	5.0	10.7	6.8	14.0	13.5	49.9	2
Tonga	16.0	8.3	10.3	15.8	11.1	61.5	3
Turkmenistan	14.0	14.5	14.7	11.7	12.7	67.6	3
Tuvalu	14.0	6.0	10.6	17.6	4.8	53.0	2
Uzbekistan	13.0	11.8	15.9	9.7	11.7	62.1	3
Vanuatu	8.0	9.8	7.9	13.0	11.2	49.9	2
Viet Nam	10.0	9.5	10.9	14.1	15.4	59.9	2

KD = key dimension, KD1 = rural household water security, KD2 = economic water security, KD3 = urban water security, KD4 = environmental water security, KD5 = water-related disaster security, NWSI = National Water Security Index.

Notes:
1. N.A. means not applicable for KD1 because Hong Kong, China; Nauru; and Singapore are 100% urban states. For these ADB members the NWSI has been determined by multiplying the sum of KD2, KD3, KD4, and KD5 by a factor of 5/4.
2. Red font means estimate, as data were not available.

Source: Asian Development Bank.

Comparing AWDO 2020, 2016, and 2013 Results

In the regional analysis (Appendix 2), AWDO 2020 results are compared with those of AWDO 2016 and AWDO 2013. AWDO 2020 presents more or less the situation in 2018, AWDO 2016 the situation in 2014, and AWDO 2013 the situation in 2009. Thus, the comparison between the three AWDOs describes the progress made in a 4- to 5-year period.

Some caution should be exercised in looking at the differences between AWDO 2020 and earlier published AWDO versions. The methodology of some of the KDs and some data sources have changed, making AWDO 2020 not directly comparable with previously published AWDO versions. The scores for 2013 and 2016 as included in this document are recalculated based on the new 2020 methodology to make the results for 2013, 2016, and 2020 comparable.

Figure A1 shows the number of ADB members and the population for 2013, 2016, and 2020 according to the five development NWS stages.

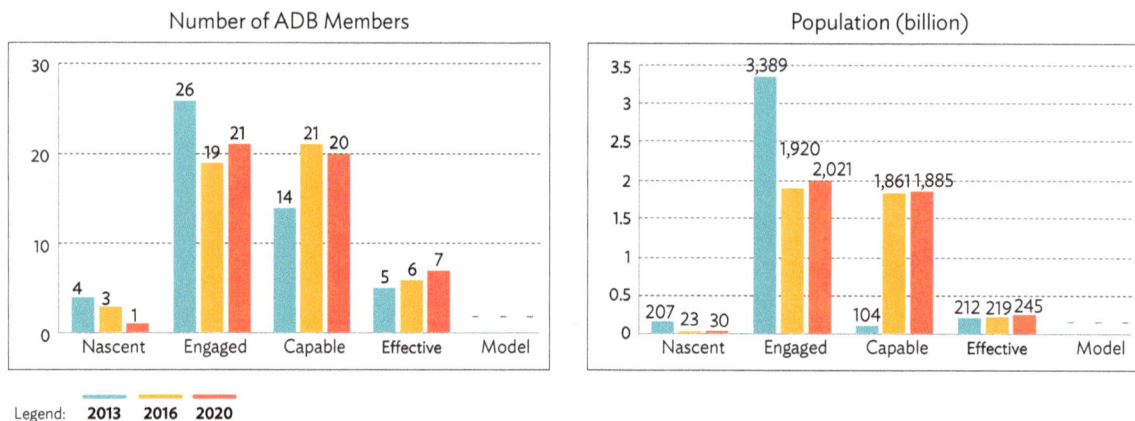

Figure A1: Number of ADB Members and People
in Five National Water Security Stages

Legend: 2013 2016 2020

Source: Asian Development Bank.

APPENDIX 2
Regional Analysis

The score calculations are done and will be presented at the country level. For presentation and comparison purposes, regional summaries will be provided. The regions identified follow the divisions of ADB (Table A2).

The total population considered in AWDO 2020 is 4,180 million (2018 population data). Note that Pakistan is included in Central and West Asia and not in South Asia. Note also that in contrast to AWDO 2016, Timor-Leste is now in Southeast Asia and not in the Pacific. The regional results are population-weighted averages. This means that the result of East Asia, for example, is very much determined by the score of the People's Republic of China (PRC), and the result of South Asia (to a somewhat lesser extent) by the score of India. The Pacific contains only 0.3% of the total population.

Population-Weighted Regional Averages

The regional average results presented in this section are population weighted. The reason for this population weighting is that a normal average in, e.g., the region East Asia (the PRC; Mongolia; and Taipei,China) does not make sense as in that case both economies have the same weight. The PRC's population is 1,395 million people while Mongolia has only 3 million. On the other hand, using a population averaging makes the average of East Asia actually equal to the score of the PRC and that score is not representative for Mongolia. For this reason, the tables in the appendix include the populated-weighted averages with and without the dominant economies. This is the case for all six regions.

Table A2: Regional Populations, 2018

Region	ADB Members	Total Population (million)	Population (%)
Central and West Asia	Afghanistan, Armenia, Azerbaijan, Georgia, Kazakhstan, Kyrgyz Republic, Pakistan, Tajikistan, Turkmenistan, and Uzbekistan	332	7.9
East Asia	People's Republic of China; Mongolia; and Taipei,China	1,422	34.0
Pacific	Cook Islands, Federated States of Micronesia, Fiji, Kiribati, Marshall Islands, Nauru, Niue, Palau, Papua New Guinea, Samoa, Solomon Islands, Tonga, Tuvalu, and Vanuatu	11	0.3
South Asia	Bangladesh, Bhutan, India, Maldives, Nepal, and Sri Lanka	1,549	37.1
Southeast Asia	Cambodia, Indonesia, Lao People's Democratic Republic, Malaysia, Myanmar, Philippines, Timor-Leste, Thailand, and Viet Nam	644	15.4
Advanced Economies	Australia; Brunei Darussalam; Hong Kong, China; Japan; New Zealand; Republic of Korea; and Singapore	222	5.3
Total Asia and the Pacific		4,180	100.0

Source: Asian Development Bank.

- Central and West Asia: Pakistan with 64% of the total regional population
- East Asia: the PRC with 98% of the total regional population
- Pacific: Papua New Guinea with 79% of the total regional population
- South Asia: India with 86% of the total regional population
- Southeast Asia: Indonesia with 41% of the total regional population
- Advanced Economies: Japan with 57% of the total regional population

This is illustrated in Figure A2.1. The figures in this document show the population-weighted averages with all ADB members in the region.

Figure A2.2 shows the population-weighted regional scores for 2013, 2016, and 2020. The line represents the average score for 2020 in which the Advanced Economies are not taken into account. As mentioned, the results of each region are very much influenced by the scores of the dominant economies. Without the dominant economies, the following are the regional NWS scores for 2020:

- Central and West Asia without Pakistan: 59.3, up 10.2 points
- East Asia without the PRC: 78.4, up 5.6 points, due to the high score of Taipei,China
- Pacific without Papua New Guinea: 54.7 points, up 9.3 points
- South Asia without India: 53.5, up 5.8 points
- Southeast Asia without Indonesia: 61.3, up 0.1 points
- Advanced Economies without Japan: 84.7, down 1.8 points

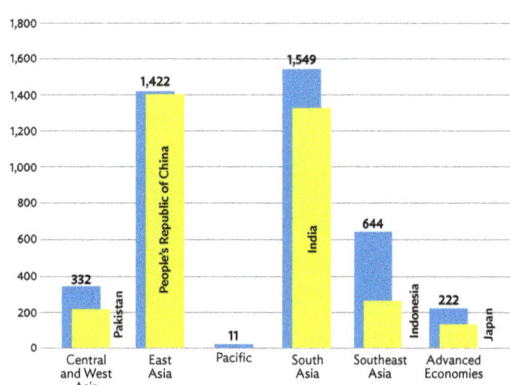

Figure A2.1: Population by Region, with Dominant Economies
(million)

Source: Asian Development Bank.

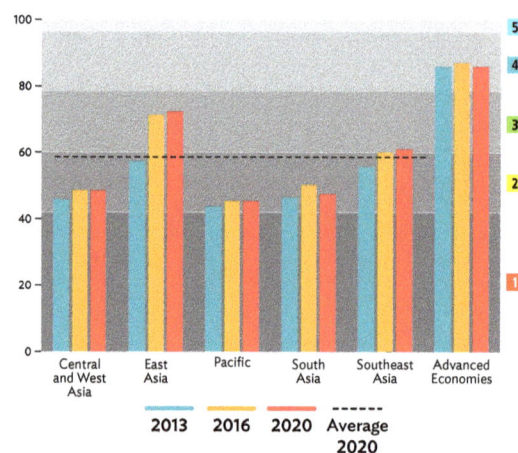

Figure A2.2: National Water Security Score by Region

2013   2016   2020   ------ Average 2020

Source: Asian Development Bank.

National Water Security Scores by Region

Central and West Asia

	Population in '000	NWS Score		
		2013	2016	2020
Afghanistan	30,075	39.2	40.0	39.5
Armenia	2,971	65.3	72.0	71.8
Azerbaijan	9,940	63.5	66.8	65.1
Georgia	3,727	60.3	62.8	64.4
Kazakhstan	18,276	66.5	72.6	73.7
Kyrgyz Republic	6,257	62.5	68.3	72.6
Pakistan	212,820	40.8	42.5	42.7
Tajikistan	9,026	49.5	54.7	58.1
Turkmenistan	5,851	62.5	66.4	67.6
Uzbekistan	32,955	55.7	64.2	62.1
Average population weighted	331,898	46.1	48.9	48.6
Average without Pakistan	119,078	55.7	60.3	59.3

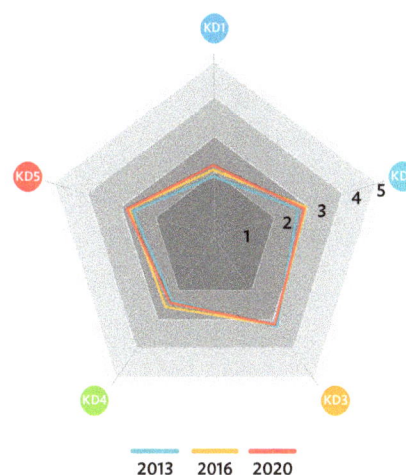

2013 2016 2020

AWDO 2020

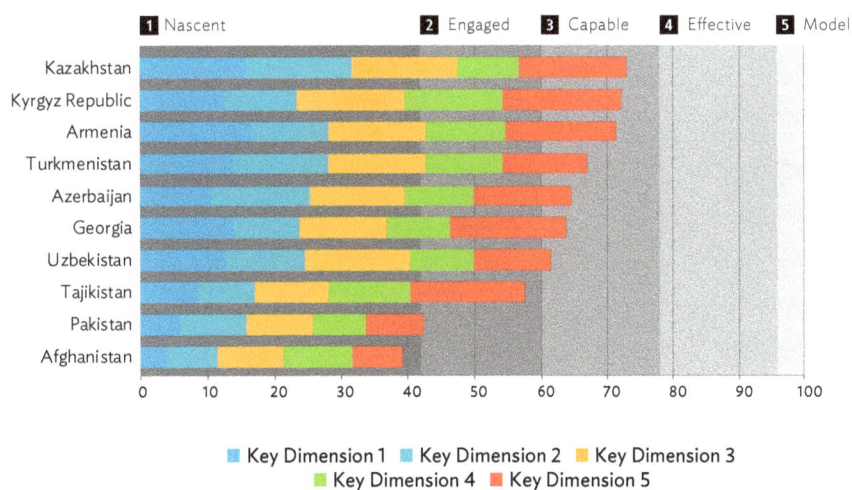

1 Nascent   2 Engaged   3 Capable   4 Effective   5 Model

Kazakhstan
Kyrgyz Republic
Armenia
Turkmenistan
Azerbaijan
Georgia
Uzbekistan
Tajikistan
Pakistan
Afghanistan

■ Key Dimension 1 ■ Key Dimension 2 ■ Key Dimension 3
■ Key Dimension 4 ■ Key Dimension 5

Source: Asian Development Bank.

East Asia

	Population in '000	NWS Score 2013	2016	2020
China, People's Republic of (PRC)	1,395,380	57.1	71.5	72.7
Mongolia	3,208	55.4	62.6	61.1
Taipei,China	23,580	72.5	76.0	80.8
Average population weighted	1,422,168	57.3	71.5	72.8
Average without the PRC	26,788	70.8	74.5	78.4

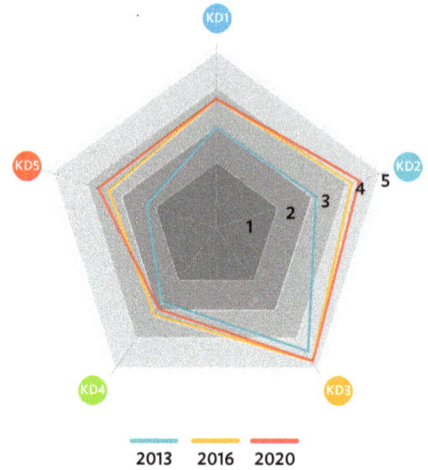

2013 2016 2020

AWDO 2020

1 Nascent   2 Engaged   3 Capable   4 Effective   5 Model

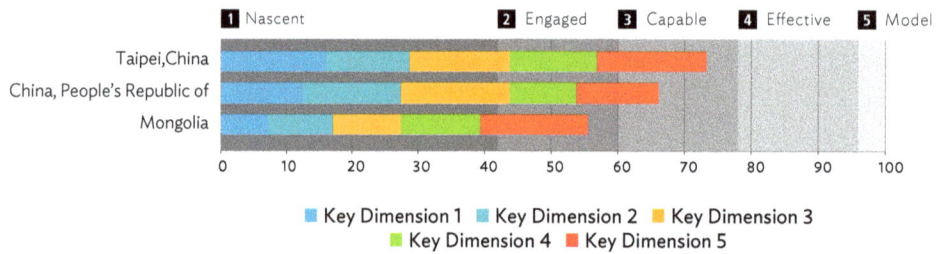

■ Key Dimension 1 ■ Key Dimension 2 ■ Key Dimension 3
■ Key Dimension 4 ■ Key Dimension 5

Source: Asian Development Bank.

Pacific

	Population in '000	NWS Score 2013	NWS Score 2016	NWS Score 2020
Cook Islands	19	66.3	70.4	72.5
Fiji	886	57.1	59.8	59.5
Kiribati	113	45.8	45.8	48.2
Marshall Islands	55	42.9	40.9	48.9
Micronesia, Federated States of	103	39.5	37.7	42.0
Nauru	11	55.9	62.0	58.6
Niue	1.7	55.5	59.9	61.0
Palau	18	62.8	69.4	73.0
Papua New Guinea (PNG)	9,019	41.1	42.8	42.8
Samoa	198	57.2	62.9	62.8
Solomon Islands	667	51.9	49.6	49.3
Tonga	100	61.4	61.4	61.5
Tuvalu	12	47.1	48.3	53.0
Vanuatu	285	51.1	49.7	49.9
Average population weighted	11,485	44.1	45.5	45.4
Average without PNG	2,466	55.3	55.3	54.7

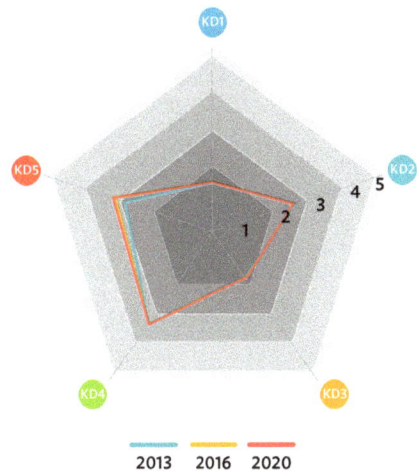

2013 2016 2020

AWDO 2020

1 Nascent   2 Engaged   3 Capable   4 Effective   5 Model

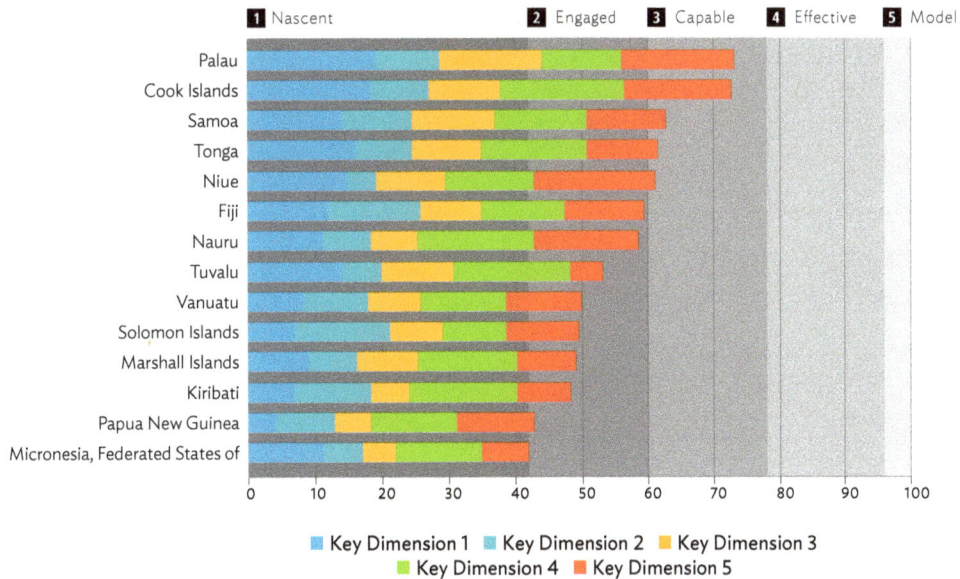

Palau
Cook Islands
Samoa
Tonga
Niue
Fiji
Nauru
Tuvalu
Vanuatu
Solomon Islands
Marshall Islands
Kiribati
Papua New Guinea
Micronesia, Federated States of

0   10   20   30   40   50   60   70   80   90   100

■ Key Dimension 1   ■ Key Dimension 2   ■ Key Dimension 3
■ Key Dimension 4   ■ Key Dimension 5

Source: Asian Development Bank.

South Asia

	Population in '000	NWS Score 2013	NWS Score 2016	NWS Score 2020
Bangladesh	164,600	47.0	51.6	52.8
Bhutan	734	60.1	61.3	62.8
India	1,332,000	46.3	50.0	46.8
Maldives	512	53.2	60.4	62.8
Nepal	29,102	50.7	53.2	52.3
Sri Lanka	21,670	56.4	58.5	60.0
Average population weighted	1,548,618	46.6	50.3	47.7
Average without India	216,618	48.5	52.6	53.5

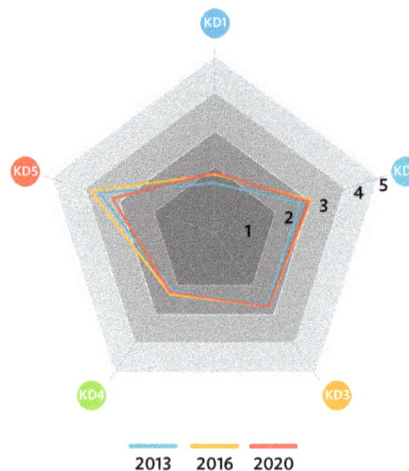

2013　2016　2020

AWDO 2020

| 1 | Nascent | 2 | Engaged | 3 | Capable | 4 | Effective | 5 | Model |

Maldives
Bhutan
Sri Lanka
Bangladesh
Nepal
India

0　10　20　30　40　50　60　70　80　90　100

■ Key Dimension 1　■ Key Dimension 2　■ Key Dimension 3
■ Key Dimension 4　■ Key Dimension 5

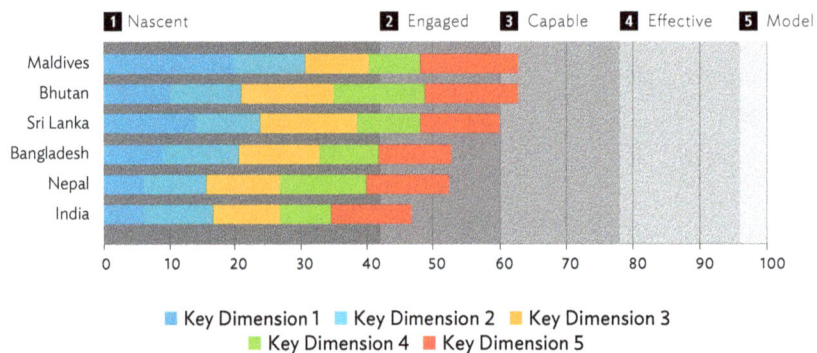

Source: Asian Development Bank.

Southeast Asia

	Population in '000	NWS Score		
		2013	2016	2020
Cambodia	15,643	55.7	58.6	57.5
Indonesia	265,000	56.6	60.5	61.0
Lao People's Democratic Republic	6,779	49.0	55.6	55.2
Malaysia	32,400	72.6	73.3	74.7
Myanmar	53,860	44.7	47.1	48.6
Philippines	106,600	59.0	67.0	67.8
Thailand	67,831	54.5	56.8	58.6
Timor-Leste	1,324	49.2	53.1	49.9
Viet Nam	94,700	54.4	58.5	59.9
Average population weighted	644,137	56.1	60.3	61.2
Average without Indonesia	379,137	55.7	60.1	61.3

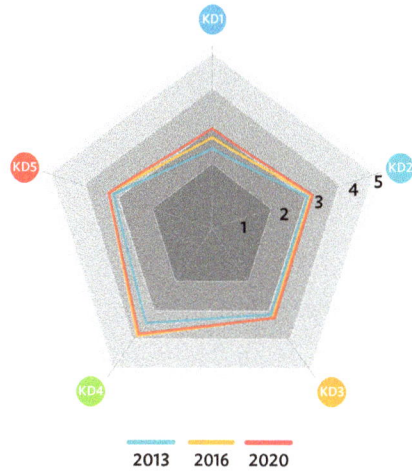

2013   2016   2020

AWDO 2020

1 Nascent   2 Engaged   3 Capable   4 Effective   5 Model

Malaysia
Philippines
Indonesia
Viet Nam
Thailand
Cambodia
Lao PDR
Timor-Leste
Myanmar

0   10   20   30   40   50   60   70   80   90   100

■ Key Dimension 1   ■ Key Dimension 2   ■ Key Dimension 3
■ Key Dimension 4   ■ Key Dimension 5

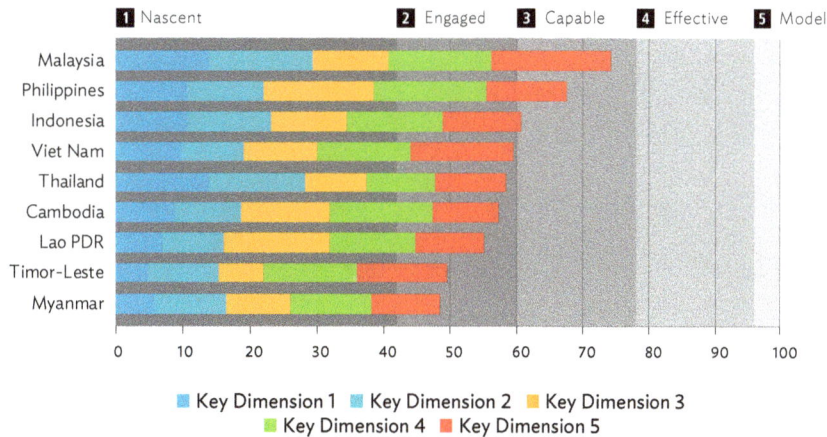

Source: Asian Development Bank.

Advanced Economies

	Population in '000	NWS Score 2013	NWS Score 2016	NWS Score 2020
Australia	24,993	87.6	88.3	87.8
Brunei Darussalam	442	74.3	74.2	75.5
Hong Kong, China	7,451	78.3	82.5	81.5
Japan	126,529	87.1	88.4	87.9
Korea, Republic of	51,607	85.9	85.1	84.0
New Zealand	4,886	88.7	88.9	89.1
Singapore	5,639	77.0	79.4	78.0
Average population weighted	221,546	86.3	87.2	86.5
Average without Japan	95,017	85.4	85.6	84.7

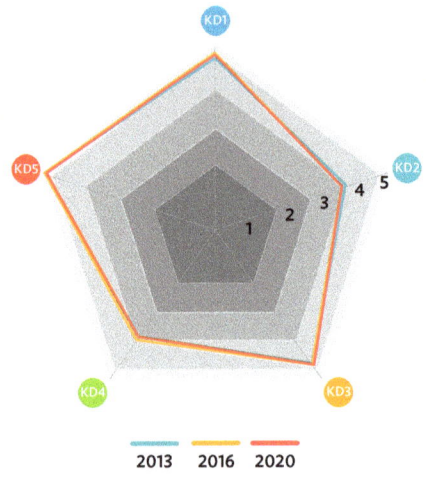

2013 2016 2020

AWDO 2020

1 Nascent   2 Engaged   3 Capable   4 Effective   5 Model

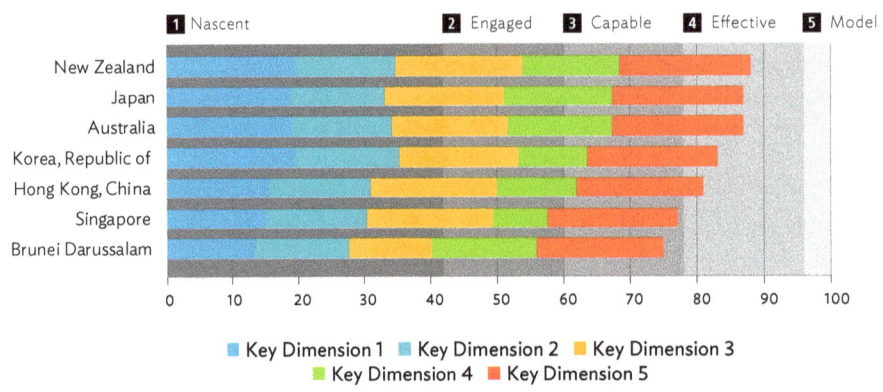

New Zealand
Japan
Australia
Korea, Republic of
Hong Kong, China
Singapore
Brunei Darussalam

0 10 20 30 40 50 60 70 80 90 100

■ Key Dimension 1   ■ Key Dimension 2   ■ Key Dimension 3
■ Key Dimension 4   ■ Key Dimension 5

Source: Asian Development Bank.

Key Dimension 1: Rural Household Water Security

Indicators

A risk framework with four indicators has been developed to create the overall key dimension (KD) 1 (rural household water security) score and measure the following:

- Indicator 1: Access to water supply—the percentage of rural people with access to different levels of water supply
- Indicator 2: Access to sanitation—the percentage of rural people with access to different levels of sanitation services
- Indicator 3: Health impacts—disability-adjusted life years (DALYs) for the impacts of water, sanitation, and hygiene (WASH) services
- Indicator 4: Affordability—the percentage of household consumption needed to afford safely managed WASH services

Indicators 1 and 2 use the Joint Monitoring Programme (JMP) service ladders and data to determine the raw score, which was then banded to find the indicator score. The following weighting was applied to take the service levels into account:

- The proportion of the rural population with at least basic service was multiplied by three.
- The proportion of the rural population with an improved service was multiplied by two.
- The proportion of the rural population with an unimproved service was kept as it was (no multiplication).

Indicator 3 uses an existing data set showing each Asian Development Bank (ADB) country's WASH-attributable disease burden measured in DALYs.

Indicator 4 uses the World Bank's Water and Sanitation Program and database with cost estimates of safely managed water services in ADB members.

The Cost Performance Index and currency exchange factors are applied to calculate these costs. Affordability is calculated as the percentage of costs of water services to total household consumption.

Changes in Methodology Compared with AWDO 2016

In the Asian Water Development Outlook (AWDO) 2020, KD1 has been redefined from household water security in 2016 to only rural household water security in 2020. Therefore, the following changes were made to the four indicators:

- Indicator 1 now targets *rural households only* and takes a more holistic perspective than solely piped water supply by considering multiple service levels.
- Indicator 2 now targets *rural households only* and takes a more holistic perspective than solely improved sanitation by considering multiple service levels.
- Indicator 3 has been refined to consider only the disease burden attributable to WASH.
- Indicator 4 has been introduced to consider affordability.

Scoring Methodology

The applied scoring methodology for KD1 is described in detail in the AWDO 2020 methodology and data report, with the following main characteristics:

- Each indicator is scored from 1 to 5.
- The KD1 score is the sum of the four indicators, with a maximum score of 20.
- The KD1 index is determined based on the banding provided in Table A3.1.

Detailed scores for KD1 are shown in Table A3.2.

The International WaterCentre, Griffith University created and populated the KD1 index, as used in AWDO 2020. The KD1 indicators were originally developed for AWDO 2013 by the United Nations Economic and Social Commission for Asia and the Pacific. The AWDO 2016 application, including an update of the methodology around the DALY parameter, was completed by the Asia Pacific Center for Water Security in Beijing.

Table A3.1: Banding Applied for Key Dimension 1 (Rural Household Water Security) Index

Index	Stage	Score	Description
5	Model	> 19.2	All rural households have access to at least basic water supply and sanitation. WASH-attributable disease burden is minimal. Water services are cheap.
4	Effective	15.6–19.2	The vast majority of rural households have access to at least basic water supply and sanitation. WASH-attributable disease burden is low. Water services are cheap.
3	Capable	12.0–15.6	A significant majority of rural households have access to at least basic water supply and sanitation. WASH-attributable disease burden is moderate. Water services are affordable.
2	Engaged	8.4–12.0	A majority of rural households have access to at least basic water supply, but a significant number of households have no access to basic sanitation. WASH-attributable disease burden is high. Water services are affordable.
1	Nascent	< 8.4	A significant number of households have no access to basic water supply or sanitation. WASH-attributable disease burden is very high. Water services are unaffordable.

WASH = water, sanitation, and hygiene.
Source: Asian Development Bank.

Table A3.2: Detailed Scores for Key Dimension 1 (Rural Household Water Security)

Economy		Piped Water	Sanitation	DALY	Affordability	Score	Index
	Scale	1–5	1–5	1–5	1–5	1–20	1–5
Afghanistan		1	1	1	1	4	1
Armenia		5	2	5	5	17	4
Australia		5	5	5	4	19	4
Azerbaijan		2	3	3	3	11	2
Bangladesh		3	1	2	3	9	2
Bhutan		4	2	3	1	10	2
Brunei Darussalam		5	3	5	1	14	3
Cambodia		2	1	2	4	9	2
China, People's Republic of		2	2	5	5	14	3
Cook Islands		5	4	4	5	18	4
Fiji		3	4	2	3	12	3
Georgia		3	2	5	4	14	3
Hong Kong, China		N.A.	N.A.	N.A.	N.A.	N.A.	N.A.
India		3	1	1	1	6	1
Indonesia		2	2	3	4	11	2
Japan		5	5	4	5	19	4

continued on next page

Table A3.2 continued

Economy	Piped Water	Sanitation	DALY	Affordability	Score	Index
Scale	1–5	1–5	1–5	1–5	1–20	1–5
Kazakhstan	3	5	4	4	16	4
Kiribati	2	1	1	3	7	1
Korea, Republic of	5	5	5	5	20	5
Kyrgyz Republic	2	5	3	3	13	3
Lao People's Democratic Republic	2	1	1	3	7	1
Malaysia	3	4	3	4	14	3
Maldives	5	5	5	5	20	5
Marshall Islands	3	1	1	4	9	2
Micronesia, Federated States of	2	3	2	4	11	2
Mongolia	1	1	3	3	8	2
Myanmar	2	1	2	1	6	1
Nauru	N.A.	N.A.	N.A.	N.A.	N.A.	N.A.
Nepal	3	1	1	1	6	1
New Zealand	5	5	5	5	20	5
Niue	4	4	2	5	15	3
Pakistan	3	1	1	1	6	1
Palau	5	5	4	5	19	4
Papua New Guinea	1	1	1	1	4	1
Philippines	3	2	2	4	11	2
Samoa	3	4	3	4	14	3
Singapore	N.A.	N.A.	N.A.	N.A.	N.A.	N.A.
Solomon Islands	1	1	2	3	7	1
Sri Lanka	3	4	4	3	14	3
Taipei,China	...	...	...	...	17.7	4
Tajikistan	2	4	2	1	9	2
Thailand	5	4	3	2	14	3
Timor-Leste	2	1	1	1	5	1
Tonga	5	3	3	5	16	4
Turkmenistan	4	5	2	3	14	3
Tuvalu	5	2	3	4	14	3
Uzbekistan	3	5	4	1	13	3
Vanuatu	2	1	1	4	8	2
Viet Nam	3	2	3	2	10	2

... = not available, DALY = disability-adjusted life year, N.A. = not applicable.

Notes:

1. It is N.A. for Hong Kong, China; Nauru; and Singapore because they are 100% urban states.
2. The score for Taipei,China (in red) is an estimate based on KD1–gross national income relation.

Source: Asian Development Bank.

APPENDIX 4
Key Dimension 2: Economic Water Security

Indicators

Economic water security is based on the performance of four indicators (one general and three specific sector indicators):

- Indicator 1: Broad economy—describes the general water-related boundary conditions for the use of water for economic purposes, including data on supply reliability, water stress, storage (dam capacity), and data availability
- Indicator 2: Agriculture—indicates water productivity, self-sufficiency, and nutrient security
- Indicator 3: Energy—indicates water productivity, self-sufficiency, and energy security
- Indicator 4: Industry—indicates water productivity, self-sufficiency, and industry security

Appendix 8 shows an overview of these indicators, as well as the units applied, data sources, data years, and data references. Further information is provided in the methodology and data report of the Asian Water Development Outlook (AWDO) 2020.

Changes in Methodology Compared with AWDO 2016

The methodology used in AWDO 2020 is similar to AWDO 2016, with the following additional elements:

- Environmental flows are taken into account in the water stress calculation.

- Nutrient security is included in the agriculture indicator.
- Self-sufficiency is included in the energy indicator.
- Self-sufficiency and industrial water security are included in the industry indicator.

These changes in the methodology are described in detail in the methodology and data report of AWDO 2020.

Scoring Methodology

The applied scoring methodology for key dimension 2 (KD2) or economic water security is described in detail in the AWDO 2020 methodology and data report. The main characteristics of the scoring methodology are the following:

- Each indicator is scored from 1 to 5.
- The KD2 score is the sum of the four indicators, with a maximum score of 20.
- The KD2 index is determined based on the banding provided in Table A4.1.

Detailed scores for KD2 are shown in Table A4.2.

The International Water Management Institute developed and populate the KD2 index, as used in AWDO 2020, AWDO 2016, and AWDO 2013. The Food and Agriculture Organization of the United Nations (FAO) was involved in developing the AWDO 2013 version.

Table A4.1: Banding Applied for Key Dimension 2 (Economic Water Security) Index

Index	Stage	Score	Description
5	Model	> 19.2	Economic water security potential is very high, meeting almost all of the following criteria: * Sufficient water is available to sustainably meet all demands, including environmental requirements. * Sufficient storage and infrastructure are available to meet demands reliably and reduce flood and drought losses to acceptable levels. * The agricultural, energy, and industrial sectors are all using water efficiently and productively. * The country is secure and self-sufficient in its production of goods, so no additional water is required for self-sufficiency. * Monitoring is performed regularly, and data and information are available to assess and adjust management when needed.
4	Effective	15.6–19.2	Economic water security potential is high, with many criteria being met.
3	Capable	12.0–15.6	Economic water security potential is moderate, with only some criteria being met.
2	Engaged	8.4–12.0	Economic water security potential is low, with a few criteria being met.
1	Nascent	< 8.4	Economic water security potential is very low, with very few, if any, criteria being met.

Source: Asian Development Bank.

Table A4.2: Detailed Scores for Key Dimension 2 (Economic Water Security)

Economy	Broad Economy	Agriculture	Energy	Industry	Score	Index
Scale	1–5	1–5	1–5	1–5	1–20	1–5
Afghanistan	2.2	1.5	1.5	2.3	7.5	2
Armenia	2.6	3.3	3.3	2.3	11.6	3
Australia	3.4	3.7	4.0	4.7	15.7	4
Azerbaijan	3.4	3.3	4.0	3.7	14.4	3
Bangladesh	1.8	3.7	3.3	3.0	11.8	3
Bhutan	2.2	1.0	5.0	3.0	11.2	2
Brunei Darussalam	2.3	2.3	4.7	5.0	14.3	3
Cambodia	2.0	3.0	1.7	3.3	10.0	2
China, People's Republic of	3.1	4.3	4.0	5.0	16.4	4
Cook Islands	2.5	...	2.0	...	9.0	2
Fiji	3.6	4.5	2.5	3.0	13.6	3
Georgia	3.0	2.7	2.3	2.0	10.0	2
Hong Kong, China	1.5	5.0	4.3	4.5	15.3	4
India	1.9	3.3	2.7	3.0	10.9	2
Indonesia	2.6	3.7	3.0	3.0	12.2	3
Japan	2.3	2.7	4.3	5.0	14.3	3
Kazakhstan	3.8	3.7	4.3	4.3	16.1	4
Kiribati	2.8	5.0	2.5	1.0	11.3	3

continued on next page

Table A4.2 continued

Economy	Broad Economy	Agriculture	Energy	Industry	Score	Index
Scale	1–5	1–5	1–5	1–5	1–20	1–5
Korea, Republic of	2.6	3.3	4.7	5.0	15.6	4
Kyrgyz Republic	3.1	3.0	2.7	2.0	10.8	2
Lao People's Democratic Republic	2.1	3.7	1.0	2.5	9.2	2
Malaysia	3.2	3.7	4.0	4.7	15.5	4
Maldives	2.7	2.5	2.5	3.0	10.7	2
Marshall Islands	2.3	...	2.0	1.0	7.0	1
Micronesia, Federated States of	1.0	...	2.5	1.0	6.0	1
Mongolia	2.6	1.7	3.3	3.3	10.9	2
Myanmar	2.6	3.3	2.3	2.5	10.8	2
Nauru	1.3	...	3.0	1.0	7.0	1
Nepal	2.0	3.3	1.7	2.7	9.7	2
New Zealand	3.6	3.3	3.7	4.3	14.9	3
Niue	1.0	...	...	...	4.0	1
Pakistan	1.8	3.0	2.7	2.5	9.9	2
Palau	1.8	...	4.5	1.0	9.7	2
Papua New Guinea	2.8	1.0	2.5	2.5	8.8	2
Philippines	2.3	3.7	3.0	2.3	11.3	3
Samoa	2.3	4.5	2.5	1.0	10.3	2
Singapore	1.9	...	4.7	5.0	15.4	4
Solomon Islands	2.5	4.0	2.5	5.0	14.0	3
Sri Lanka	1.9	3.0	2.7	2.5	10.0	2
Taipei,China	1.0	5.0	4.7	...	14.2	3
Tajikistan	2.6	2.0	2.3	1.5	8.4	2
Thailand	2.7	3.7	3.7	4.7	14.7	3
Timor-Leste	2.0	2.0	3.0	3.7	10.7	2
Tonga	2.8	...	2.5	1.0	8.3	2
Turkmenistan	1.6	3.7	4.7	4.5	14.5	3
Tuvalu	2.0	...	...	1.0	6.0	1
Uzbekistan	2.1	4.0	3.0	2.7	11.8	3
Vanuatu	2.3	4.5	2.0	1.0	9.8	2
Viet Nam	1.8	3.7	1.5	2.5	9.5	2

... = not available.

Source: Asian Development Bank.

Key Dimension 3: Urban Water Security

Indicators

Consistent with the Sustainable Development Goals (SDGs) for water and sanitation and the new scope, key dimension 3 (KD3) or urban water security now measures the extent to which ADB members provide safely managed and affordable water and sanitation services for their urban communities to achieve desired outcomes sustainably. The KD3 method aligns with the KD1 method, where appropriate. The KD3 score is composed of the following indicators:

- Indicator 1: Water supply—percentage of population in each service ladder
- Indicator 2: Sanitation—percentage of population in each service ladder
- Indicator 3: Affordability—percentage of costs/expenditure
- Indicator 4: Drainage—percentage of economic damage due to floods and storms/gross domestic product (GDP)
- Indicator 5: Environment—taken from KD4 (environmental water security)

Although the KDs seem independent, the indicators used to quantify them are not completely unconnected. For example, indicators 4 and 5 are also used in KD4 and KD5 (water-related disaster security). Indicators 4 and 5 have been included in KD3 to reflect the full picture of urban water security, taking into account urban water quality and urban water flooding. Giving a maximum score of 1 to the weights of indicators of 4 and 5 (compared with a maximum score of 5 to indicators 1–3) avoids too much overlap between KDs and indicators.

Changes in Methodology Compared with AWDO 2016

The 2020 methodology is closely aligned with the 2016 methodology, with the following key enhancements:

- The inputs and scoring criteria for indicator 1 (water supply) and indicator 2 (sanitation, previously wastewater) have been reshaped to align with improved data sets.
- Indicator 3 (affordability) has been included.
- The scoring for indicator 4 (drainage) has been aligned with the scoring of KD5.
- Indicators 4 (drainage) and 5 (environment) have both been given a maximum score of 1.
- The urban growth factor has been removed from the assessment of current KD3 and integrated into a new risk indicator.

Indicators 1 and 2 were aligned with the SDGs and improved data sets that distinguish the service ladder of the WHO/UNICEF Joint Monitoring Programme. For countries that did not report data yet for the highest service level of "safely managed" supply and sanitation, this might result in an underestimation of the scores of the indicators for these countries. This is for example the case for Thailand which scored "only" 2 (out of 5) for both supply and sanitation while the actual situation on urban water supply (54% cover for piped and 46% for non-piped supply on premises) and sanitation (12% cover for sewer connections, 87% for septic tanks) indicate good performance.

The score for India might be underestimated as the results of recent programs on KD3 are not included yet in the data used for the Asian Water Development Outlook (AWDO) 2020. See Box 3 for an overview of these programs.

The methodology of KD3 now also includes an analysis of the risk to future urban water security. This risk is calculated for ADB members where data are available for the following elements:

- urban growth—percentage per annum,
- total water consumption—liters per person per day,
- nonrevenue water—"difference between water supplied and water sold as a percentage of total water supplied," and
- energy costs—percentage of annual operating costs.

This risk analysis is not included in this document. Reference is made to the KD3 report for a more detailed description of this analysis and the results.

As data become available, the risk assessment may be expanded to cover all ADB members in the future, including other risk elements.

Scoring Methodology

The applied scoring methodology for KD3 is described in detail in the AWDO 2020 methodology and data report. The main characteristics of the scoring methodology are the following:

- Indicators 1–3 (water supply, sanitation, and affordability) are scored from 1 to 5, according to a unique set of scoring criteria.
- Indicator 4 (drainage) and river health indicator are scored from 0 to 1. A score of 0 indicates poor conditions, while 1 indicates good conditions (no constraint on urban water security).
- The five indicators are summed with a maximum score of 17.
- The sum of KD3 is multiplied by a factor 20/17 to make KD3 comparable with other KDs.
- The KD3 index is determined based on the banding provided in Table A5.1.

Detailed scores for KD3 are shown in Table A5.2.

For more details on the data quality assessment (low, moderate, and high), refer to the full KD3 report.

The International WaterCentre and Advanced Water Management Centre, both in Australia, developed the KD3 index. They populated the index and analyzed the KD3 results. The International WaterCentre did that for the two previous AWDO versions (2013 and 2016).

Table A5.1: Banding Applied for Key Dimension 3 (Urban Water Security) Index

Index	Stage	Score	Description
5	Model	> 19.2	A very high proportion of the urban population receives affordable, safely managed water and sanitation services, with low economic impacts of floods and storms, and high environmental water security.
4	Effective	15.6–19.2	A high proportion of the urban population receives affordable, safely managed water and sanitation services, with low economic impacts of floods and storms, and high environmental water security.
3	Capable	12.0–15.6	A moderate proportion of the urban population receives affordable, safely managed water and sanitation services, with acceptable economic impacts of floods and storms, and acceptable environmental water security.
2	Engaged	8.4–12.0	A low proportion of the urban population receives affordable, safely managed water and sanitation services, with potentially high economic impacts of floods and storms, and low environmental water security.
1	Nascent	< 8.4	A very low proportion of the urban population receives affordable, safely managed water and sanitation services, with potentially high economic impacts of floods and storms, and low environmental water security.

Source: Asian Development Bank.

Table A5.2: Detailed Scores for Key Dimension 3 (Urban Water Security)

Economy	Water Supply	Sanitation	Affordability	Drainage	Environment	Sum	Adjusted Score	Index
Scale	1–5	1–5	1–5	0–1	0–1	1–17	1–20	1–5
Afghanistan	2	1	4	1.00	0.50	8.50	10.0	2
Armenia	2	4	5	1.00	0.50	12.50	14.7	3
Australia	5	5	4	0.25	0.75	15.00	17.6	4
Azerbaijan	2	5	4	1.00	0.50	12.50	14.7	3
Bangladesh	4	1	4	0.75	0.50	10.25	12.1	3
Bhutan	4	1	5	1.00	0.75	11.75	13.8	3
Brunei Darussalam	2	2	5	1.00	0.75	10.75	12.6	3
Cambodia	4	2	4	0.50	0.75	11.25	13.2	3
China, People's Republic of	5	5	4	0.75	0.50	15.25	17.9	4
Cook Islands	2	2	3	1.00	1.00	9.00	10.6	2
Fiji	2	2	3	0.00	0.75	7.75	9.1	2
Georgia	3	3	4	0.75	0.50	11.25	13.2	3
Hong Kong, China	5	5	5	1.00	0.50	16.50	19.4	5
India	2	1	4	0.75	0.50	8.25	9.7	2
Indonesia	2	1	5	1.00	0.75	9.75	11.5	3
Japan	5	5	4	0.75	1.00	15.75	18.5	4
Kazakhstan	2	5	5	1.00	0.50	13.50	15.9	4
Kiribati	1	1	1	1.00	1.00	5.00	5.9	1
Korea, Republic of	5	5	4	1.00	0.50	15.50	18.2	4
Kyrgyz Republic	5	2	5	1.00	0.75	13.75	16.2	4
Lao People's Democratic Republic	3	5	4	0.75	0.75	13.50	15.9	4
Malaysia	2	2	4	1.00	0.75	9.75	11.5	3
Maldives	2	2	3	1.00	0.25	8.25	9.7	2
Marshall Islands	1	2	3	1.00	0.75	7.75	9.1	2
Micronesia, Federated States of	1	1	1	0.25	0.75	4.00	4.7	1
Mongolia	2	1	5	1.00	0.75	9.75	11.5	3
Myanmar	2	1	4	0.50	0.75	8.25	9.7	2
Nauru	2	1	1	1.00	1.00	6.00	7.1	1
Nepal	3	1	4	0.75	0.75	9.50	11.2	2
New Zealand	5	5	5	0.75	0.75	16.50	19.4	5
Niue	2	2	3	1.00	0.75	8.75	10.3	2
Pakistan	3	1	4	0.50	0.25	8.75	10.3	2
Palau	3	3	5	1.00	1.00	13.00	15.3	4
Papua New Guinea	1	1	1	1.00	0.75	4.75	5.6	1
Philippines	5	3	5	0.25	1.00	14.25	16.8	4
Samoa	3	4	3	0.00	0.75	10.75	12.6	3
Singapore	5	5	5	1.00	0.25	16.25	19.1	4

continued on next page

Table A5.2 continued

Economy		Water Supply	Sanitation	Affordability	Drainage	Environment	Sum	Adjusted Score	Index
	Scale	1–5	1–5	1–5	0–1	0–1	1–17	1–20	1–5
Solomon Islands		2	1	3	0.25	0.50	6.75	7.9	2
Sri Lanka		5	2	4	0.75	0.50	12.25	14.4	3
Taipei,China		...	...	...	...	0.75	14.01	16.5	4
Tajikistan		2	2	4	0.75	0.75	9.50	11.2	2
Thailand		2	2	3	0.25	0.50	7.75	9.1	2
Timor-Leste		2	1	1	1.00	0.75	5.75	6.8	1
Tonga		2	2	4	0.00	0.75	8.75	10.3	2
Turkmenistan		5	2	4	1.00	0.50	12.50	14.7	3
Tuvalu		4	1	2	1.00	1.00	9.00	10.6	2
Uzbekistan		5	3	4	1.00	0.50	13.50	15.9	4
Vanuatu		2	1	3	0.00	0.75	6.75	7.9	2
Viet Nam		2	2	4	0.50	0.75	9.25	10.9	2

... = not available.

Note: The score for Taipei,China (in red) is an estimate based on KD3 (urban water security)–gross national income relation.

Source: Asian Development Bank.

APPENDIX 6
Key Dimension 4: Environmental Water Security

Indicators

Key dimension 4 (KD4) or environmental water security uses two indicators:

- Indicator 1: Catchment and Aquatic System Condition Index (CASCI)
- Indicator 2: Environmental Governance Index (EGI)

Catchment and Aquatic System Condition Index. The methodology for CASCI relies on a large-scale, spatially explicit data collection of variables that determine the health of aquatic ecosystems and the state of key outcome variables. The approach is based on the Driver-Pressure-State-Impact-Response framework,[1] which describes the interaction between society and the environment:

- Pressures
  - » Riparian land cover change
  - » Hydrological alteration
  - » Groundwater depletion
- States
  - » Water quality
  - » Riverine connectivity

Environmental Governance Index. The EGI quantifies the results of (good) governance, rather than a metric of governance itself. It relies on the following country-level metrics collated by the Yale Environmental Performance Index for 2018:

- Wastewater treatment
- Terrestrial protected areas
- Sustainable nitrogen management index

KD4 explicitly accounts for several Sustainable Development Goals (SDGs) that are relevant to aquatic ecosystem health (apart from SDG 6). Given the inherent relationship between the landscape through which a river flows and its condition, the KD4 indicators and sub-indicators are also relevant to targets within SDGs 3, 14, and 15, taking into account the connection between catchment management, aquatic ecosystem health, the availability of safe water, and coastal ecosystem conditions. Taking steps to improve each element of KD4 will assist a country in reaching many of the SDG targets.

Changes in Methodology Compared with AWDO 2016

The methodology to determine KD4 has changed substantially for Asian Water Development Outlook (AWDO) 2020 compared with AWDO 2016. The River Health Index used in AWDO 2016 was replaced by CASCI, which also includes the flow regulation indicator used in AWDO 2013. The EGI has remained more or less like in 2016. In AWDO 2016, EGI included a country-level summary of total forest cover change, but this was removed for 2020 because the same underlying data source is used as part of the riparian land cover change data included in CASCI.

It is noted that some parts of the methodology for KD4 is less applicable for urban and small-island states, e.g., Singapore. This explains partly some of the low scores of ADB members.

---

[1] It is a causal framework describing the interactions between society and the environment adopted by the European Environment Agency. It is an extension of the pressure-state-response model originally developed by the OECD.

Scoring Methodology

The applied scoring methodology for KD4 is described in detail in the AWDO 2020 methodology and data report. The main characteristics of the scoring methodology are the following:

- Both indicators are scored from 1 to 5 using a hierarchical aggregation approach.
- The two indicators are summed with a maximum score of 10.

- The sum of KD4 score is multiplied by a factor 2 to make KD4 comparable with other KDs.
- The KD4 index is determined based on the banding provided in Table A6.1.

Detailed scores for KD4 are shown in Table A6.2.

The KD4 methodology has been developed by the International WaterCentre, who also determined the KD4 scores and analyzed KD4 results for all AWDO versions (2013, 2016, and 2020).

Table A6.1: Banding Applied for Key Dimension 4 (Environmental Water Security) Index

Index	Stage	Score	Description
5	Model	> 19.2	Very good outcomes of environmental governance and very limited pressures on aquatic ecosystems
4	Effective	15.6–19.2	Moderate to good outcomes of environmental governance and limited pressures on aquatic ecosystems
3	Capable	12.0–15.6	Potentially moderate outcomes of environmental governance and pressures on aquatic ecosystems, or very good outcomes in one and poor performance in the other
2	Engaged	8.4–12.0	Moderate to poor outcomes of environmental governance and severe pressures on aquatic ecosystems
1	Nascent	< 8.4	Poor outcomes of environmental governance and significant pressures on aquatic ecosystems

Source: Asian Development Bank.

Table A6.2: Detailed Scores for Key Dimension 4 (Environmental Water Security)

Economy		CASCI	EGI	Sum	Score	Index
	Scale	1–5	1–5	1–10	1–20	1–5
Afghanistan		2.6	2.7	5.3	10.6	2
Armenia		1.7	4.2	6.0	11.9	2
Australia		3.5	4.2	7.7	15.4	3
Azerbaijan		2.0	3.1	5.0	10.1	2
Bangladesh		2.2	2.3	4.5	9.0	2
Bhutan		3.5	3.5	6.9	13.9	3
Brunei Darussalam		3.9	3.8	7.7	15.5	3
Cambodia		3.5	4.2	7.7	15.4	3
China, People's Republic of		2.0	3.5	5.4	10.8	2
Cook Islands		4.7	...	4.7	18.9	4
Fiji		4.1	1.9	6.1	12.1	3

continued on next page

Table A6.2 continued

Economy	Scale	CASCI 1–5	EGI 1–5	Sum 1–10	Score 1–20	Index 1–5
Georgia		2.8	1.9	4.7	9.5	2
Hong Kong, China		3.0	...	3.0	12.0	3
India		1.7	2.3	4.0	8.1	1
Indonesia		4.1	3.1	7.2	14.4	3
Japan		3.3	5.0	8.3	16.5	4
Kazakhstan		2.4	2.3	4.7	9.4	2
Kiribati		4.6	3.5	8.1	16.2	4
Korea, Republic of		1.7	3.5	5.2	10.4	2
Kyrgyz Republic		3.3	4.2	7.5	15.0	3
Lao People's Democratic Republic		3.0	3.5	6.5	13.0	3
Malaysia		3.5	4.2	7.7	15.4	3
Maldives		...	1.9	1.9	7.7	1
Marshall Islands		3.8	...	3.8	15.0	3
Micronesia, Federated States of		4.6	1.9	6.5	13.0	3
Mongolia		3.9	2.7	6.6	13.2	3
Myanmar		3.5	2.7	6.2	12.3	2
Nauru		4.4	...	4.4	17.5	4
Nepal		3.5	3.1	6.6	13.1	3
New Zealand		3.9	3.5	7.4	14.7	3
Niue		3.3	...	3.3	13.3	3
Pakistan		1.5	2.3	3.8	7.7	1
Palau		4.6	...	4.6	18.3	3
Papua New Guinea		4.3	1.9	6.3	12.5	3
Philippines		4.1	4.2	8.4	16.7	4
Samoa		5.0	1.9	6.9	13.8	3
Singapore		2.0	1.9	3.9	7.8	1
Solomon Islands		3.6	1.2	4.8	9.5	2
Sri Lanka		2.2	2.7	4.9	9.7	2
Taipei,China		2.8	4.2	7.1	14.1	3
Tajikistan		2.6	3.5	6.1	12.1	3
Thailand		2.0	3.1	5.0	10.1	2
Timor-Leste		3.9	3.1	7.0	14.0	3
Tonga		4.4	3.5	7.9	15.8	4
Turkmenistan		2.4	3.5	5.9	11.7	2
Tuvalu		4.4	...	4.4	17.5	4
Uzbekistan		2.2	2.7	4.9	9.7	2
Vanuatu		4.6	1.9	6.5	13.0	3
Viet Nam		2.8	4.2	7.1	14.1	3

... = not available, CASCI = Catchment and Aquatic System Condition Index, EGI = Environmental Governance Index.

Note: In case the score of an indicator was not available, the value of the other indicator has been used to determine the overall key dimension 4 (KD4) score.

Source: Asian Development Bank.

Key Dimension 5: Water-Related Disaster Security

Indicators

Key dimension 5 (KD5) is a national assessment of risk to water-related disasters. It comprises three indicators, which describe the risk categories of water-related disasters:

- Indicator 1: Climatological risk, including drought
- Indicator 2: Hydrological risk, including coastal, riverine and flash flooding, and mudslides
- Indicator 3: Meteorological risk, including tropical storms, convection storms, and associated storm surge

Based on the definition of risk by the United Nations Office for Disaster Risk Reduction and the international academic community, these risks are determined by the following three sub-indicators of water-related disaster risk:

- **Hazard exposure**—the situation of people in areas prone to a process or phenomenon that may cause loss of life, injury, or other health impacts; social and economic disruption; or environmental degradation
    - » Using historical data from the Emergency Events Database (EM-DAT), this indicator is based on the total number of people affected by water-related disasters in a nation over the last decade.
- **Vulnerability**—the conditions determined by physical, social, economic, and environmental factors or processes that increase the susceptibility of an individual, community, or system to the impacts of hazards

    - » Population susceptibilities are described at a national level, including poverty, infant mortality, and dependency ratios.
    - » Quality of governance and disaster financing is assessed based on corruption perception, national adherence to international accords on disaster risk, and development assistance levels.
    - » Environmental degradation is described by rates of deforestation and unsustainable extraction of groundwater resources.
- **Capacity**—the strengths, attributes, and resources available within an organization, community, or society to manage and reduce disaster risks and strengthen resilience
    - » Hard coping capacities are described based on the assessments of key infrastructure related to irrigation, reservoir capacity, and information and communication technology systems.
    - » Soft coping capacities are described based on the national rates of basic educational attainment and literacy, as well as savings and earning capacities.

Based on the definition of disaster risk by the United Nations Office for Disaster Risk Reduction, these sub-indicators are used to quantify water-related disaster risk using the following formula:

$$R = (HE \times V \times (1-C))^{1/3}$$

where;

R = *Disaster Risk Index* (0–1)

HE = *Hazard – Exposure Sub-indicator* (0–1)

V = *Vulnerability Sub-indicator* (0–1)

C = *Capacity Sub-indicator* (0–1)

Changes in Methodology Compared with AWDO 2016

Significant changes have been made to the KD5 assessment since 2016, primarily by focusing on risk rather than resilience. This was done by incorporating data on the actual hazard impacts, rather than assessing hazard exposure, as was done in 2016. This fundamental change resulted in major shifts to KD5 scores. Thus, KD5 scores in 2020 cannot be directly compared with published KD5 scores in 2016 and 2013. However, due to the relationship between disaster risk and resilience, much of the assessment's basic structure has been retained, including vulnerability and capacity assessment. One exception was the incorporation of a sub-sub-indicator based on the extent to which nations enacted recommendations from the Hyogo Framework for Action 2005–2015, which broadens KD5 to include assessments of national disaster risk financing, insurance markets, and disaster risk reduction policy.

Scoring Methodology

The applied scoring methodology for KD5 is described in detail in the Asian Water Development Outlook (AWDO) 2020 methodology and data report. The main characteristics of the scoring methodology are the following:

- Each indicator is scored from 1 to 5.
- The three indicators are summed with a maximum score of 15.
- The sum of the KD5 score is multiplied by 20/15 (resulting in a maximum score of 20) to make KD5 comparable with the other KDs.

The KD5 index is determined based on the banding provided in Table A7.1. Detailed scores for KD5 are shown in Table A7.2.

The KD5 approach was developed in 2013 by the International Centre for Water Hazard and Risk Management in Japan and was reformulated in 2020 by Korea Institute of Civil Engineering and Construction Technology, representing a shift to disaster risk assessment.

Table A7.1: Banding Applied for Key Dimension 5 (Water-Related Disaster Security) Index

Index	Stage	Score	Description
5	Model	> 19.2	Full achievement with sustained commitment to disaster risk reduction
4	Effective	15.6–19.2	Systematic commitment at policy level to disaster risk reduction
3	Capable	12.0–15.6	Institutional commitment to disaster risk reduction
2	Engaged	8.4–12.0	Minor progress achieved in disaster risk reduction
1	Nascent	< 8.4	No progress made in disaster risk reduction, or progress stopped or moved backward

Source: Asian Development Bank.

Table A7.2: Detailed Scores for Key Dimension 5 (Water-Related Disaster Security)

Economy	Climatological (Droughts)	Hydrological (Floods)	Meteorological (Storms)	Total	Score	Index
Scale	1–5	1–5	1–5	1–15	1–20	1–5
Afghanistan	1.0	1.0	3.5	5.5	7.3	1
Armenia	4.2	4.2	4.1	12.4	16.6	4
Australia	5.0	5.0	5.0	15.0	20.0	5
Azerbaijan	2.7	4.2	4.3	11.2	14.9	3
Bangladesh	2.9	1.7	3.6	8.2	11.0	2
Bhutan	3.0	3.0	4.5	10.4	13.9	3
Brunei Darussalam	4.7	4.6	5.0	14.3	19.1	4
Cambodia	2.2	1.0	4.2	7.4	9.8	2
China, People's Republic of	3.5	2.2	4.5	10.2	13.5	3
Cook Islands	3.7	4.5	3.8	12.0	16.0	4
Fiji	3.5	4.0	1.7	9.2	12.3	3
Georgia	4.7	3.8	4.8	13.3	17.7	4
Hong Kong, China	4.5	4.6	5.0	14.1	18.8	4
India	2.0	2.8	4.3	9.1	12.1	3
Indonesia	2.9	2.4	3.7	8.9	11.9	2
Japan	5.0	4.7	5.0	14.7	19.6	5
Kazakhstan	2.8	4.8	4.7	12.2	16.3	4
Kiribati	1.0	1.4	3.5	6.0	7.9	1
Korea, Republic of	5.0	4.8	5.0	14.8	19.7	5
Kyrgyz Republic	4.4	4.4	4.4	13.2	17.6	4
Lao People's Democratic Republic	2.0	2.3	3.4	7.6	10.1	2
Malaysia	4.1	4.6	5.0	13.7	18.2	4
Maldives	3.5	3.5	4.1	11.1	14.8	3
Marshall Islands	1.0	1.6	3.9	6.5	8.6	2
Micronesia, Federated States of	1.8	2.7	1.0	5.5	7.4	1
Mongolia	4.4	4.2	4.5	13.1	17.5	4
Myanmar	3.5	1.1	3.0	7.7	10.2	2
Nauru	3.5	4.2	4.1	11.8	15.7	4
Nepal	2.5	2.5	4.2	9.3	12.4	3
New Zealand	5.0	5.0	5.0	15.0	20.0	5
Niue	4.4	4.6	4.8	13.8	18.4	4
Pakistan	1.5	1.1	4.0	6.6	8.8	2
Palau	3.8	4.6	4.2	12.7	16.9	4
Papua New Guinea	2.5	2.7	3.8	9.0	12.0	2
Philippines	4.4	2.4	2.2	9.0	12.1	3
Samoa	2.7	2.7	3.6	9.0	12.0	3

continued on next page

Table A7.2 continued

Economy	Climatological (Droughts)	Hydrological (Floods)	Meteorological (Storms)	Total	Score	Index
Scale	1–5	1–5	1–5	1–15	1–20	1–5
Singapore	5.0	5.0	5.0	15.0	20.0	5
Solomon Islands	4.0	1.3	2.9	8.2	10.9	2
Sri Lanka	2.7	1.6	4.5	8.8	11.8	2
Taipei,China	4.6	4.3	4.8	13.7	18.3	4
Tajikistan	4.4	4.3	4.4	13.0	17.4	4
Thailand	2.4	1.2	4.4	8.1	10.8	2
Timor-Leste	2.4	3.8	3.9	10.1	13.5	3
Tonga	3.1	4.2	1.0	8.3	11.1	2
Turkmenistan	1.0	4.1	4.4	9.5	12.7	3
Tuvalu	1.0	1.6	1.0	3.6	4.8	1
Uzbekistan	1.3	3.0	4.4	8.8	11.7	2
Vanuatu	3.3	4.1	1.0	8.4	11.2	2
Viet Nam	4.3	3.3	4.0	11.6	15.4	3

Source: Asian Development Bank.

Overview of Databases Used for Indicators and Sub-Indicators

Indicator/ Sub-Indicator	Measure	Unit	Data Years	Coverage ADB Members	Database / Reference	Comments
KD1 (Key Dimension 1: Rural Household Water Security)						
Water supply and water sanitation	Proportion of people with access to basic services. Higher service levels are given higher scores.	Calculate composite score from 0 to 3	2017	45/45	Joint Monitoring Programme (JMP 2019)	Data are disaggregated between urban and rural areas. Very good reference for water, sanitation, and hygiene (WASH). No gender delineation.
Health	Disability-adjusted life years (DALYs) attributable to poor WASH	DALYs/person	2016	40	Burden of disease from inadequate WASH for selected adverse health outcomes (Prüss-Ustün et al. 2019)	Uses disease burden and mortality estimates data set as raw data (WHO 2018). No urban–rural delineation. Has estimates for 2016 only.
Affordability	Affordability of safely managed water supply, sanitation, and hygiene services	Cost of safely managed water services as a % of household consumption	2015	40	Costs of meeting the 2030 Sustainable Development Goal (SDG) targets (Prüss-Ustün et al. 2019)	Builds up an overall cost per person of having access to safely managed WASH for both urban and rural areas.
General			2018	45	ADB Key Indicators Database (ADB 2019)	This database was used to source Household Final Consumption, Cost Performance Index, currency exchange rates, and population.
KD2 (Key Dimension 2: Economic Water Security)						
Broad Economy						
Water scarcity and water stress	Total freshwater withdrawal/total renewable water resources (TRWR)		2017	10–37	Food and Agriculture Organization of the United Nations (FAO) AQUASTAT, satellite remote sensing analysis	TRWR is available for 37 ADB members. Withdrawal data availability varied by sector and country. Although 2017 data were used, they may not have been updated from previous years. Remote sensing analysis filled some gaps in agricultural water withdrawal. Environmental flow requirements were considered a withdrawal.

continued on next page

Table continued

Indicator/ Sub-Indicator	Measure	Unit	Data Years	Coverage ADB Members	Database / Reference	Comments
Reliability and self-sufficiency Intra-annual precipitation variability	Average coefficient of variation of monthly rainfall within years for the previous 20 years.		2018	43	Climatic Research Unit Time-Series version 4.03 (CRU TS4.03)	20 years of data up to 2018 were used
Interannual precipitation variability	Coefficient of variation of annual rainfall over the previous 20 years		2018	43	CRU TS4.03	
Storage capacity	Reservoir storage/ TRWR		2017	17	FAO AQUASTAT	Used data from the last year reported for each country.
Resilience and security Storage drought duration index	Total dam capacity/ total freshwater withdrawal per month/mean annual drought duration			34	Eriyagama, Smakhtin, and Gamage (2009); New et al. (2002); FAO AQUASTAT	
Governance / data availability					All data sources	Availability of 18 data points were assessed: Time series precipitation, Total renewable water resources, Reservoir capacity, Industrial water withdrawal, Municipal water withdrawal, Agricultural water withdrawal, Environmental flow requirements, Population, Total gross domestic product (GDP), Agricultural GDP, Agricultural import value, Agricultural export value, Food Supply, Industrial GDP, Export of goods, Import of goods, Electricity production, and Electricity consumption.
Agriculture						
Water productivity	Total agricultural production/total agricultural water depletion	$ million/ cubic kilometers (km³)	2016–2018	32	Remote sensing analysis; World Bank	
Self-sufficiency	Ratio of water used in consumption to water used in production of agricultural goods		2016–2018	42	ADB; Hoekstra and Mekonnen (2012); World Bank	This is the net virtual water import in agriculture.

continued on next page

Table continued

Indicator/ Sub-Indicator	Measure	Unit	Data Years	Coverage ADB Members	Database / Reference	Comments
Nutrient security	Additional calories/ capita needed to reach the Asia and Pacific average consumption	Kilocalorie (kcal)/ capita/ day	2016–2017	37	FAOSTAT	
Energy						
Water productivity	Gigawatt-hour (GWh) production/ water consumption	GWh/km3		30	Gerbens-Leenes, Hoekstra, and Meer (2008); IPCC (2012); IEA; Mekonnen, Gerbens-Leenes, and Hoekstra (2015)	A country's diverse energy sources were linked to global consumption averages to determine the water consumption associated with energy production in a country.
Self-sufficiency	Electricity consumption/ electricity production		2016–2018	43	ADB	
Energy security	Additional installed capacity needed to raise per capita electricity production/ Asia and Pacific average		2016–2018	44	ADB	
Industry						
Water productivity	Industrial GDP/ industrial withdrawal	$ million/km^3	2016–2018	10	FAO AQUASTAT; World Bank	Industrial water withdrawal data were limiting.
Self-sufficiency	Trade deficit in industrial goods	$	2016–2018	29	ADB; FAOSTAT; World Bank	Data on the imports and exports of industrial goods were limiting.
Industry security	Additional industrial output needed to raise per capita industrial GDP to the Asia and Pacific average	$	2016–2018	39	ADB; World Bank	
KD3 (Key Dimension 3: Urban Water Security)						
Water supply	Water supply service level for urban for safely managed, basic, limited, unimproved, and surface water	%	41 urban 49 national	2013, 2016, 2017	JMP 2019	
Sanitation	Sanitation service level for urban coverage for safely managed, basic, limited, unimproved, and open defecation	%	41 urban 48 national	2013, 2016, 2017	JMP 2019	

continued on next page

Table continued

Indicator/ Sub-Indicator	Measure	Unit	Data Years	Coverage ADB Members	Database/ Reference	Comments
Affordability	Equivalent annual costs for urban safely managed water and sanitation services	$/capita	41	2015	Hutton and Varughese (2016)	
	Consumer prices indices		43	2013, 2016, 2017	ADB Key Indicators Database	
	Household final consumption expenditures	% of GDP	37	2013, 2016, 2017	ADB Key Indicators Database	
	GDP, purchasing power parity (PPP)	Current international $/capita	48	2013, 2016, 2017	ADB Key Indicators Database	
	Consumptive household expenditure	Current international $/capita	8	2010–2016	Household Income and Expenditure Survey (HIES) reports	
Drainage	Flood, storm damages	$	45	1999–2013 2002–2016 2005–2019	Emergency Events Database (EM-DAT) 2019	EM-DAT is a global disaster database maintained by the Centre for Research on the Epidemiology of Disasters
	Population	No.	48	2013, 2016, 2017	JMP 2019	
	GDP, PPP	Current international $/capita	48	2013, 2016, 2017	ADB Key Indicators Database	
Environment	Indicator provided by KD4					
KD4 (Key Dimension 4: Environmental Water Security)						
Catchment and Aquatic System Health						
Riparian land cover change	Forest cover loss Grass and shrub cover loss Maximum extent of inundations	% of riparian land altered over time	2000–2018	38	Hansen et al. (2013); Climate Change Initiative – Land Cover (ESA CCI-LC) project; UCL Geomatics (2017); Pekel et al. (2016)	Data for riparian land cover change come from a range of sources as they include the alteration of forests, shrubs, and grasslands within a 100 meter buffer around and including the maximum extent of rivers and wetlands.
Hydrological alteration	Extent of flow alteration relative to an unmodified flow regime	Weighted proportion	2005–2017	33	Wisser et al. 2010	Using the TerraClimate, a high-resolution data set of monthly climate forcings (Abatzoglou et al. 2018).

continued on next page

Table continued

Indicator/Sub-Indicator	Measure	Unit	Data Years	Coverage ADB Members	Database/Reference	Comments
Groundwater depletion	Long-term decline in aquifer levels	Trend in change over time	2002–2016	33	Gravity Recovery and Climate Experiment (GRACE)	For Pacific island nations that are too small to be detected by the GRACE satellite, an assessment of groundwater resources from Geoscience Australia was used (Stewart et al. 2014).
Water quality	Pollution loads Sustainable Nitrogen Management Index (SNMI)	Weighted score (0–1)	Around 2000 and 2014	37	Vörösmarty et al. (2010); Zhang and Davidson (2016)	This information captures the amount of pressure on the waterways in a given country and offsets that pressure by assessing the efficiency of agricultural use of nitrogen, as measured by the SNMI.
Riverine connectivity	Index based on remote sensing and existing large-scale spatial data set	Discharge weighted average %	2018	42	Grill et al. (2019)	Index was weighted with total annual discharge, which means a country is penalized more for low levels of connectivity in rivers with higher discharge.
Environmental governance	Yale Environmental Performance Index assessment from 2018				Wendling et al. (2018)	
Wastewater treatment	% wastewater treated	%	2018	42	United Nations Statistics Division Organisation for Economic Co-operation and Development (OECD) Pinsent Masons Water Yearbook FAO AQUASTAT	Data from these sources were supplemented by data from national reports
Terrestrial protected area	% country's biomes in terrestrial protected areas	%	2018	42	World Database on Protected Areas (WDPA) World Resources Institute's "Terrestrial Ecoregions of the World" data set	
Sustainable nitrogen management	Efficiency of nitrogen use (SNMI)	Unitless (0–100)	2016	42	Zhang and Davidson (2016)	

continued on next page

Table continued

Indicator/ Sub-Indicator	Measure	Unit	Data Years	Coverage ADB Members	Database/ Reference	Comments
KD5 (Key Dimension 5: Water-Related Disaster Security)						
Hazard-Exposure	No. of people	No.	2010–2019		EM-DAT	
Vulnerability						
Lack of governance	Corruption Perceptions Index (CPI)	Index	2018		Transparency international	Transparency International publishes the CPI annually, which ranks countries based on how corrupt their public sectors are seen to be.
Lack of national disaster risk reduction	Hyogo Framework for Action progress reports	Index	2007–2015		National governments	
Environmental vulnerability	Forested land cover Freshwater withdrawal as % of renewable water	% %	2010–2016 2013–2017		UNSTAT AQUASTAT	
Aid dependency	Official development assistance (ODA) as % of GDP	%	2017		World Bank	
Vulnerable populations	Population below $1.9 per day Population age > 65 years Population age 0–14 years Infant mortality	% % %	2012–2017 2018 2018 2017		World Bank UNSTAT UNSTAT UNSTAT	
Capacity						
Economic capacity	GDP per capita Gross domestic savings as % of GDP	$/capita %	2018 2017		World Bank World Bank	
Reservoir capacity	Reservoir capacity	CuM/capita	2019		Global Reservoir and Dam (GRandD)	The GRandD database is the result of an effort made by the scientific community called the Global Water System Project and managed by McGill University.
Information and communication technology (ICT) capacity	Mobile subscriptions Internet subscriptions	% %	2017 2017		UNSTAT UNSTAT	

continued on next page

Table continued

Indicator/ Sub-Indicator	Measure	Unit	Data Years	Coverage ADB Members	Database / Reference	Comments
Educational capacity	Literacy rate	%	2012–2017		UNSTAT	
	Primary school enrolment rate	%	2010–2017		UNSTAT	
Agricultural infrastructure	Land equipped for irrigation/total	%	2016		FAOSTAT	
Other	Population		2018		UNSTAT	

Database Sources and References:

Abatzoglou, J. T. et al. 2018. TerraClimate, a High-Resolution Global Dataset of Monthly Climate and Climatic Water Balance from 1958–2015. *Scientific Data*. 5 (170191).

ADB. Key Indicators Database (accessed 2 October 2020).

Eriyagama, N., V. Smakhtin, and N. Gamage. 2009. Mapping Drought Patterns and Impacts: A Global Perspective. *IWMI Research Report*. No. 133. Colombo, Sri Lanka: International Water Management Institute.

Gerbens-Leenes, P. W., A. Y. Hoekstra, and T. H. Meer. 2008. Water Footprint of Bio-Energy and Other Primary Energy Carriers. *Value of Water Research Report Series*. No. 29. Delft, the Netherlands: UNESCO-IHE Institute for Water Education.

Grill, G. et al. 2019. Mapping the World's Free-Flowing Rivers. *Nature*. 569. pp. 215–221.

Hansen, M. C. et al. 2013. High-Resolution Global Maps of 21st-Century Forest Cover Change. *Science*. 342 (6160). pp. 850–853.

Hoekstra, A. Y. and M. M. Mekonnen. 2012. The Water Footprint of Humanity. *Proceedings of the National Academy of Sciences*. 109 (9). pp. 3232–3237.

Hutton, G. and M. Varughese. 2016. *The Costs of Meeting the 2030 Sustainable Development Goal Targets on Drinking Water, Sanitation, and Hygiene*. Washington, DC: World Bank.

Intergovernmental Panel on Climate Change (IPCC). 2012. *Renewable Energy Sources and Climate Change Mitigation: Special Report of the Intergovernmental Panel on Climate Change*. New York: Cambridge University Press.

Joint Monitoring Programme. 2019. *Progress on Household Drinking Water, Sanitation and Hygiene 2000-2017. Special Focus on Inequalities*. New York: United Nations Children's Fund (UNICEF) and World Health Organization.

Mekonnen, M. M., P. W. Gerbens-Leenes, and A. Y. Hoekstra. 2015. The Consumptive Water Footprint of Electricity and Heat: A Global Assessment. Environmental Science: *Water Research & Technology*. 1 (3). pp. 285–297.

New, M. et al. 2002. A High-Resolution Data Set of Surface Climate over Global Land Areas. *Climate Research*. 21 (1). pp. 1–25.

Pekel, J. et al. 2016. High-Resolution Mapping of Global Surface Water and Its Long-Term Changes. *Nature*. 540. pp. 418–422.

Prüss-Ustün, A. et al. 2019. Burden of Disease from Inadequate Water, Sanitation and Hygiene for Selected Adverse Health Outcomes: An Updated Analysis with a Focus on Low-and Middle-Income Countries. *International Journal of Hygiene and Environmental Health*. 222 (5). pp. 765–777.

Stewart, G. et al. 2014. Pacific Island Groundwater Vulnerability to Future Climates Dataset. *Geoscience Australia, Canberra*.

UCL Geomatics. 2017. *Land Cover CCI Product User Guide Version 2.0*.

Vörösmarty, C. J. et al. 2010. Global Threats to Human Water Security and River Biodiversity. *Nature*. 467. pp. 555–561.

Wendling, Z. et al. 2018. *2018 Environmental Performance Index*. New Haven, Connecticut: Yale Center for Environmental Law & Policy.

Wisser, D. et al. 2010. Reconstructing 20th Century Global Hydrography: A Contribution to the Global Terrestrial Network-Hydrology (GTN-H). *Hydrology and Earth System Sciences*. 14. pp. 1–24.

Zhang, X. and E. Davidson. 2016. *Sustainable Nitrogen Management Index (SNMI): Methodology*. Frostburg, Maryland: University of Maryland Center for Environmental Science.

APPENDIX 9
OECD Survey on Water Governance in Asia

Governance data were collected in December 2019 and January 2020 for a total of 48 ADB members, comprising 4 Organisation for Economic Co-operation and Development (OECD) countries and 44 non-OECD economies: Afghanistan; Armenia; Australia; Azerbaijan; Bangladesh; Bhutan; Brunei Darussalam; Cambodia; the People's Republic of China; the Cook Islands; Fiji; Georgia; Hong Kong, China; India; Indonesia; Japan; Kazakhstan; Kiribati; the Republic of Korea; the Kyrgyz Republic; the Lao People's Democratic Republic; Malaysia; Maldives; the Marshall Islands; the Federated States of Micronesia; Mongolia; Myanmar; Nauru; Nepal; New Zealand; Pakistan; Palau; Papua New Guinea; the Philippines; Samoa; Singapore; Solomon Islands; Sri Lanka; Taipei,China; Tajikistan; Thailand; Timor-Leste; Tonga; Turkmenistan; Tuvalu; Uzbekistan; Vanuatu; and Viet Nam.[1]

The governance survey comprises 46 questions, distributed into 12 sections, which were answered using secondary data and information obtained from the following sources:

(i)   published reports and documents by international organizations, including the Asian Development Bank (ADB), the World Bank, the OECD, the Food and Agricultural Organization of the United Nations (FAO), the United Nations Development Programme, and the International Water Management Institute;

(ii)   Asia and the Pacific national government policies, strategies, and legislation (in English);

(iii) published reports and peer-reviewed research papers; and

(iv)   other gray literature.

The structure of the questionnaire followed the 12 OECD Water Governance principles:

- Section 1: Clear roles and responsibilities
- Section 2: Appropriate scales
- Section 3: Policy coherence
- Section 4: Capacity development
- Section 5: Data and information
- Section 6: Financing
- Section 7: Regulatory frameworks
- Section 8: Innovation
- Section 9: Integrity and transparency
- Section 10: Stakeholder engagement
- Section 11: Trade-offs
- Section 12: Monitoring and evaluation

The data from the survey were processed to (i) overview water governance characteristics in Asia, (ii) provide quantified evidence regarding governance gaps, and (iii) show the diversity of governance situations across Asia and the Pacific.

---

[1]   Niue has recently become an ADB member and could not be included in the survey yet.

Methodological Note on Financial Analysis of the OECD

This appendix documents the methods and data sources used to assess the Asian Development Bank (ADB) members' financing needs and capacities for water supply and sanitation, irrigation, and flood protection. A more comprehensive note is appended to a forthcoming publication of ADB and the Organisation for Economic Co-operation and Development (OECD).[1]

The assessment's greatest challenge is the lack of baseline data. To overcome this challenge, the OECD has made use of the data produced by the World Bank, the Global Water Intelligence, World Resources Institute, and the OECD.

Scenarios are based on the data produced by the International Institute for Applied Systems Analysis and their "Shared Socioeconomic Pathways 2: Middle of the Road" scenario, combined with the Intergovernmental Panel on Climate Change Representative Concentration Pathway 8.5.[2] For flood risk analysis, the Representative Concentration Pathway 8.5 is used (the high end of carbon dioxide equivalent parts per million projections).

Table A10.1 shows data sources for investment needs, and Table A10.2 shows data sources for financing capacity.

---

[1]   OECD and ADB. Forthcoming. Financing Water Security for Sustainable Growth in the Asia-Pacific Region.
[2]   International Institute for Applied Systems Analysis. SSP Scenario Database (accessed March 2020).

Table A10.1: Summary of Data Sources: Investment Needs

	Description of Analysis	Unit	Country Coverage	Data Sources
Water and Sanitation				
Cost of achieving SDG 6 + Safely managed connections	Model 2015–2030 annual average	$ and % GDP	39/49	Rozenberg and Fay (2019); Hutton and Varughese (2016)
Irrigation				
Cost of irrigation	Model 2015–2030 annual average	% of GDP	Subregional estimates	Rozenberg and Fay (2019)
Flood Protection				
GDP exposure of coastal flood risk with subsidence	Model projected exposure in 2030	$ and % GDP/ population exposed	44/49	World Resources Institute
GDP exposure of riverine flood risk			44/49	World Resources Institute

GDP = gross domestic product, SDG = Sustainable Development Goal.

Sources: G. Hutton and M. Varughese. 2016. *The Costs of Meeting the 2030 Sustainable Development Goal Targets on Drinking Water, Sanitation, and Hygiene.* Washington, DC: World Bank; J. Rozenberg and M. Fay. 2019. Beyond the Gap: How Countries Can Afford the Infrastructure They Need while Protecting the Planet. *Sustainable Infrastructure.* Washington, DC: World Bank; and World Resources Institute. Aqueduct Global Flood Analyzer (accessed 1 September 2020).

Table A10.2: Summary of Data Sources: Financing Capacity

	Description of Analysis	Country Coverage	Data Sources
Financing Strategies			
Respective role of revenue from tariffs, public finance	Anecdotal, GLAAS 2019 report	Limited	OECD (2019); WHO and UN-Water (2019)
ODA flows	WSS, Total ODA flows, 2010–2017	41 ADB members	OECD iLibrary
Financing Options			
Experience with commercial finance	Anecdotal	25 ADB members	WHO and UN-Water (2019); ADB (2017)
Equity			
Micro affordability	GWI survey of city tariffs, World Bank Debt Sustainability database	108 cities in 20 ADB members	GWI (2019)

ADB = Asian Development Bank, GLAAS = Global Analysis and Assessment of Sanitation and Drinking-Water, GWI = Global Water Intelligence, ODA = official development assistance, OECD = Organisation for Economic Co-operation and Development, WHO = World Health Organization, WSS = water supply and sanitation.

Sources: ADB. 2017. *Meeting Asia's Infrastructure Needs.* Manila; GWI. 2020. *The Global Water Tariff Survey;* OECD. 2019. Making Blended Finance Work for Water and Sanitation: Unlocking Commercial Finance for SDG 6. *OECD Studies on Water.* Paris: OECD Publishing; OECD iLibrary. Official Development Assistance Indicators (accessed 1 September 2020); WHO and UN-Water. 2019. *National Systems to Support Drinking-Water, Sanitation and Hygiene: Global Status Report 2019.* Geneva: WHO.

www.ingramcontent.com/pod-product-compliance
Lightning Source LLC
Chambersburg PA
CBHW050043220326
41599CB00045B/7265